Building Design and Construction Systems

Building Design and Construction Systems

Picard Henderson

Larsen & Keller
www.larsen-keller.com

Building Design and Construction Systems
Picard Henderson
ISBN: 978-1-64172-120-2 (Hardback)

 Larsen & Keller

Published by Larsen and Keller Education,
5 Penn Plaza,
19th Floor,
New York, NY 10001, USA

Cataloging-in-Publication Data

Building design and construction systems / Picard Henderson.
    p. cm.
Includes bibliographical references and index.
ISBN 978-1-64172-120-2
1. Building. 2. Architectural design. 3. Buildings-- Computer-aided design.
4. Construction industry I. Henderson, Picard.
TH153 .B85 2019
690--dc23

For more information regarding Larsen and Keller Education and its products, please visit the publisher's website www.larsen-keller.com

Table of Contents

Permissions

Index

Preface

A building can be defined as a structure, which has a roof and walls. The architectural, technical and engineering applications for designing buildings is referred to as building design. Zoning requirement, investigation of the environmental impact of construction, budgeting, scheduling, logistics, inconvenience to the public, construction-site safety, etc. are some of the considerations involved in design and execution. The construction of building is carried out differently for residential and non-residential or commercial setups. It involves the principal areas of planning, design and financing. The topics included in this book on building design and construction systems are of the utmost significance and bound to provide incredible insights to readers. It explores all the important aspects of these fields in the present day scenario. It will serve as a reference to structural engineers, architects, building designers and students.

Given below is the chapter wise description of the book:

Chapter 1, A building is a structure with walls and a roof, built for residence, privacy or storage. It can be constructed in a variety of shapes and sizes. This chapter has been carefully written to provide a brief introduction to buildings and its common types. **Chapter 2,** Architecture involves the process of designing, planning and constructing buildings and other structures. This chapter has been carefully structured to provide a detailed analysis of architecture, interior design, sustainable architecture, climate-adaptive building shell, computer-aided architectural design, active design and passive design, besides others. **Chapter 3,** Construction is the process of building any infrastructure. It starts with planning, designing, financing and continues till the completion of the project. The chapter closely examines the key concepts of building construction, such as building code, construction 3D printing, steel frame and earthquake-resistant structures. **Chapter 4,** Foundation is an essential aspect of an architectural structure. It connects to the ground and allows the transference of the load of the structure to the ground. Foundations are either deep or shallow. Some of the important considerations of foundation engineering are underpinning, framing, building envelope, house raising and basement waterproofing, which have been extensively discussed in this chapter. **Chapter 5,** In architecture, the term 'structural system' refers to the load-resisting sub-system of an architectural structure. The transference of loads occurs through interconnected elements. Some of the structural elements in buildings are walls, floor, roof, ceiling, etc., which have been adequately covered in this chapter. Some of the diverse topics in this chapter also address the important facets of ventilation and plumbing system. **Chapter 6,** Modular buildings are houses or structures that consists of repeated modules or sections. Its construction is a unique process that involves the installation of the prefabricated sections at the intended site of the building. The topics elaborated in this chapter on modular construction, modular building, prefabricated building and portable building will help in developing a better perspective of modular building and construction. **Chapter 7,** Building design is the process of the application of engineering, architectural and technical principles to the design of buildings. This chapter has been written to provide a comprehensive understanding of the significant aspects of building design such as building insulation, architectural lighting, house wiring, electronic security system, light fixture and carpentry.

Indeed, my job was extremely crucial and challenging as I had to ensure that every chapter is informative and structured in a student-friendly manner. I am thankful for the support provided by my family and colleagues during the completion of this book.

Picard Henderson

Building and its Types

A building is a structure with walls and a roof, built for residence, privacy or storage. It can be constructed in a variety of shapes and sizes. This chapter has been carefully written to provide a brief introduction to buildings and its common types.

Building means any structure that has roof and walls especially a permanent structure. It can be any structure that is designed or intended for support, enclosure, shelter or protection of person, animals or property having a permanent roof that is supported by columns or walls.

High-rise Building

High-rise building, also called high-rise, multistory building tall enough to require the use of a system of mechanical vertical transportation such as elevators. The skyscraper is a very tall high-rise building.

The first high-rise buildings were constructed in the United States in the 1880s. They arose in urban areas where increased land prices and great population densities created a demand for buildings that rose vertically rather than spread horizontally, thus occupying less precious land area. High-rise buildings were made practicable by the use of steel structural frames and glass exterior sheathing. By the mid-20th century, such buildings had become a standard feature of the architectural landscape in most countries in the world.

The foundations of high-rise buildings must sometimes support very heavy gravity loads, and they usually consist of concrete piers, piles, or caissons that are sunk into the ground. Beds of solid rock are the most desirable base, but ways have been found to distribute loads evenly even on relatively soft ground. The most important factor in the design of high-rise buildings, however, is the building's need to withstand the lateral forces imposed by winds and potential earthquakes. Most high-rises have frames made of steel or steel and concrete. Their frames are constructed of columns (vertical-support members) and beams (horizontal-support members). Cross-bracing or shear walls may be used to provide a structural frame with greater lateral rigidity in order to withstand wind stresses. Even more stable frames use closely spaced columns at the building's perimeter, or they use the bundled-tube system, in which a number of framing tubes are bundled together to form exceptionally rigid columns.

High-rise buildings are enclosed by curtain walls; these are non-load-bearing sheets of glass, masonry, stone, or metal that are affixed to the building's frame through a series of vertical and horizontal members called mullions and muntins.

The principal means of vertical transport in a high-rise is the elevator. It is moved by an electric motor that raises or lowers the cab in a vertical shaft by means of wire ropes. Each elevator cab is also engaged by vertical guide tracks and has a flexible electric cable connected to it that provides power for lighting, door operation, and signal transmission.

Because of their height and their large occupant populations, high-rises require the careful provision of life-safety systems. Fire-prevention standards should be strict, and provisions for adequate means of egress in case of fire, power failure, or other accident should be provided. Although originally designed for commercial purposes, many high-rises are now planned for multiple uses. The combination of office, residential, retail, and hotel space is common.

There are a number of issues arose because of the tall buildings:

Air Pollution

There are many sources of the air pollution in the city such as cars that produce CO. With increasing the height the density of the CO will be increasing. If there is a tower, this increasing can be happening till around 6-9 floors after that this increasing will be decrease irregularly. As known that, the wind load increase with increasing the height. So, if the tower has a wind flow from above to down, the CO will be Shanghai or any other cities that known as a high rise city can not be measure as a tall building or skyscraper. It will be the high-rise building. On the other hand, the same tower with the same tall can be known as a tall building in many of the cities in Europe. separate around and make the volume of the pollution bigger.

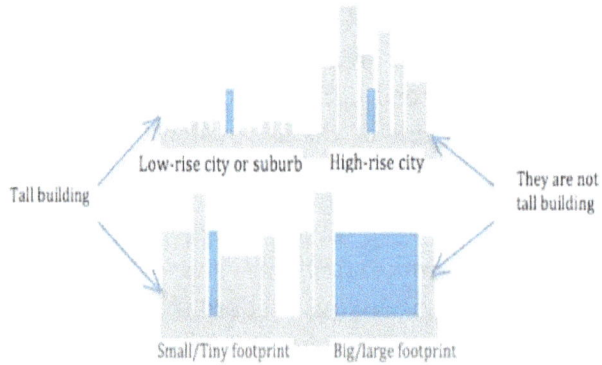

Figure: Relation between tall buildings and urban context.

Sunshine

Skyscraper can get a sun solar easily especially the upper floors because its height. What about the impacts of the building on the surroundings area, making shadow and avoiding getting sun solar directly from the sun are the appear points that many researchers conceder about it. The affects of this shadow and shading are change in different climate and block the sun with change the access, sun light and solar sun. For instance, in hot climate zone, shading the many urban space in long period time is good and helpful for daily activity. On the other hand in designing the Passivhous buildings in cold climate, it will be avoid solar gain for the buildings especially low raise buildings around the tall buildings. Also it has impacts on vegetation and green area. In agriculture, having the over shading area, it is not easy and not suitable for planting many of vegetation.as known that direct sun is very important for growing plants and green architecture. Green area and plants inside the urban is vital important case for social activities. As clear that this shading go back to a number of factors as:

- Differences in height of buildings (sky line)

- Tall buildings direction and location according to surrounding neighborhoods(with sun path and time)

- Mass and form (geometry)

- Urban space between the buildings (distance)

There are many experiments and case study have done for showing the case. The diagram figure shows the model analysis of ISTANBUL LEVENT REGION in different time and date and natural lighting.

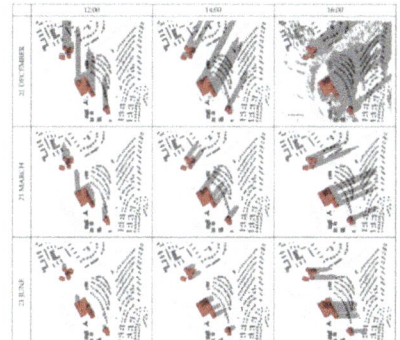      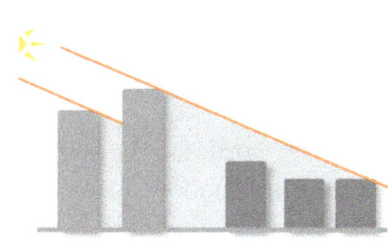

A. Istanbullevent region shadow   B. Urban Geometry.   C. Orientation and overshadowing

Wind Flow

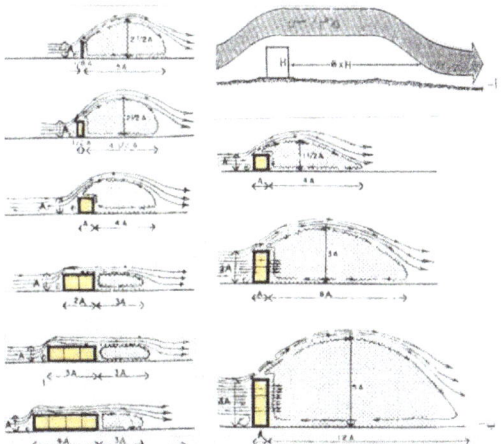

Figure: Relation between wind shadow with building height and their depth.

For the Wind flow, as the sunshine there are two different sides. The tall building can create the wind flow past the urban buildings or can avoid the airflow inside the urban planning. If the buildings are not near to each other, these impacts will be minimizing to very low level. Tall buildings can change the direction of the wind in urban planning. On the other hand, if there is quite high density with similar building height, the ventilation will be better. For air shadow, tall buildings in urban planning increase the air shadow. This shadow increases with increasing height of buildings. About the depth of the buildings not very effect able building till more than four times of building height.

In the it can be seen that the high rise buildings have a great impact on the their physical boundaries. There is a relation between the height and shadow distance that is greater than the building height about four times. To illustrate that, if the building has 20m elevations, the distance shadow will be 80m length. For the height of the shadow, it will be about one and half of the,

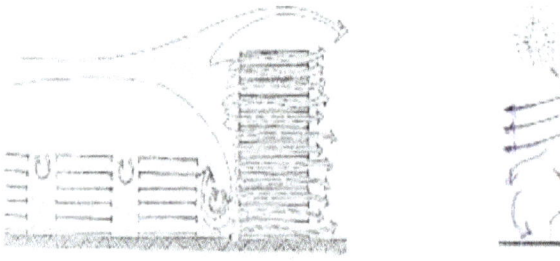

A. High-rise building enhances vortex and pollution around it
B. High-rise building prevent wind and reflect solar radiation on low-rise buildings building height.

In another word, for the same building with 20m heights, the shadow elevation will be about 30 m. This wind shadow dose not change two much with change the building depth except for those depth that more than four times of the height of the buildings. For all cases the air velocity is a vital boint for increasing and decreasing the shadow.

Plant vertical green area and roofs can solve the problem of air pollution. This green area also helpful for the building itself what help the building to create the produce clean and fresh air and decrease the temperature in the hot climate zone.

Views

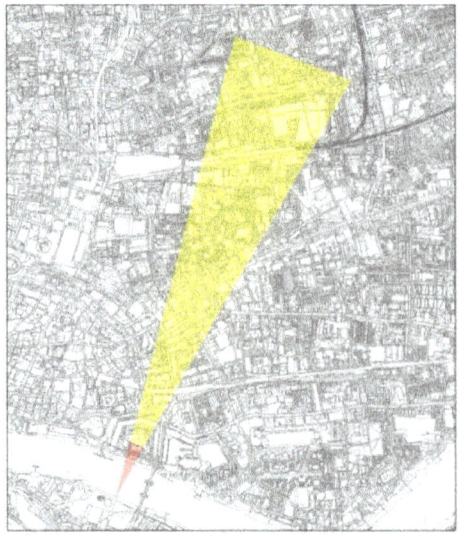

Figure: Example of protecting view of London

About the view, high-rise buildings, not like the low-rise, do block the view and visualsfrom other location of the city because their height. In many cities as London there is some rules for avoiding block views.Protected view is an important issue in urban planning especially if there is a grate global icon or historical landmark in the city and existing high raise buildings.

Low-rise Building

Low-rise residential buildings include the smallest buildings produced in large quantities. Single-family detached houses, for example, are in the walk-up range of one to three stories and typically meet their users' needs with about 90 to 180 square metres (about 1,000 to 2,000 square feet) of enclosed floor space. Other examples include the urban row house and walk-up apartment buildings. Typically these forms have relatively low unit costs because of the limited purchasing power of their owners. The demand for this type of housing has a wide geographic distribution, and therefore most are built by small local contractors using relatively few large machines (mostly for earth moving) and large amounts of manual labour at the building site. The demand for these buildings can have large local variations from year to year, and small builders can absorb these economic swings better than large organizations. The building systems developed for this market reflect its emphasis on manual labour and its low unit costs. A proportion of single-family detached houses are "factory-built"; that is, large pieces of the building are prefabricated and then transported to the site, where considerable additional work is required to complete the finished product.

Building Architecture: Designing Concepts

Architecture involves the process of designing, planning and constructing buildings and other structures. This chapter has been carefully structured to provide a detailed analysis of architecture, interior design, sustainable architecture, climate-adaptive building shell, computer-aided architectural design, active design and passive design, besides others.

Architecture

Architecture, the art and technique of designing and building, as distinguished from the skills associated with construction. The practice of architecture is employed to fulfill both practical and expressive requirements, and thus it serves both utilitarian and aesthetic ends. Although these two ends may be distinguished, they cannot be separated, and the relative weight given to each can vary widely. Because every society—whether highly developed or less so, settled or nomadic—has a spatial relationship to the natural world and to other societies, the structures they produce reveal much about their environment (including climate and weather), history, ceremonies, and artistic sensibility, as well as many aspects of daily life.

The characteristics that distinguish a work of architecture from other built structures are

1. The suitability of the work to use by human beings in general and the adaptability of it to particular human activities,

2. The stability and permanence of the work's construction, and

3. The communication of experience and ideas through its form.

All these conditions must be met in architecture. The second is a constant, while the first and third vary in relative importance according to the social function of buildings. If the function is chiefly utilitarian, as in a factory, communication is of less importance. If the function is chiefly expressive, as in a monumental tomb, utility is a minor concern. In some buildings, such as churches and city halls, utility and communication may be of equal importance.

Theory of Architecture

The term *theory of architecture* was originally simply the accepted translation of the Latin term *ratiocinatio* as used by Vitruvius, a Roman architect-engineer of the 1st century CE, to differentiate intellectual from practical knowledge in architectural education, but it has come to signify the total basis for judging the merits of buildings or building projects. Such reasoned judgments are an essential part of the architectural creative process. A building can be designed only by a continuous creative, intellectual dialectic between imagination and reason in the mind of each creator.

A variety of interpretations has been given to the term *architectural theory* by those who have written or spoken on the topic in the past. Before 1750 every comprehensive treatise or published lecture course on architecture could appropriately be described as a textbook on architectural theory. But, after the changes associated with the Industrial Revolution, the amount of architectural knowledge that could be acquired only by academic study increased to the point where a complete synthesis became virtually impossible in a single volume.

The historical evolution of architectural theory is assessable mainly from manuscripts and published treatises, from critical essays and commentaries, and from the surviving buildings of every epoch. It is thus in no way a type of historical study that can reflect accurately the spirit of each age and in this respect is similar to the history of philosophy itself. Some architectural treatises were intended to publicize novel concepts rather than to state widely accepted ideals. The most idiosyncratic theories could (and often did) exert wide and sometimes beneficial influence, but the value of these influences is not necessarily related to the extent of this acceptance.

The analysis of surviving buildings provides guidance that requires great caution, since, apart from the impossibility of determining whether or not any particular group of buildings (intact or in ruins) constitutes a reliable sample of the era, any such analyses will usually depend on preliminary evaluations of merit and will be useless unless the extent to which the function, the structure, and the detailing envisaged by the original builders can be correctly re-established. Many erudite studies of antique theories are misleading because they rest on the assumption that the original character and appearance of fragmentary ancient Greek and Hellenistic architectural environments can be adequately deduced from verbal or graphic "reconstructions." Even when buildings constructed before 1500 remain intact, the many textbooks dealing with antique and medieval theories of architecture seldom make qualitative distinctions and generally imply that all surviving antique and medieval buildings were good, if not absolutely perfect.

Nevertheless, the study of the history of architectural philosophy, like that of the history of general philosophy, not only teaches what past generations thought but can help individuals decide how they themselves should act and judge. For those desirous of establishing a viable theory of architecture for their own era, it is generally agreed that great stimulus can be found in studying historical evidence and in speculating on the ideals and achievements of those who created this evidence.

Distinction Between the History and Theory of Architecture

The distinction between the history and theory of architecture did not emerge until the mid-18th century. Indeed, the establishment of two separate academic disciplines was not even nominal until 1818, when separate professorships with these titles were established at the École des Beaux-Arts in Paris. Even then, however, the distinction was seldom scrupulously maintained by either specialist. It is impossible to discuss meaningfully the buildings of the immediate past without discussing the ideals of those who built them, just as it is impossible to discuss the ideals of bygone architects without reference to the structures they designed. Nevertheless, since any two disciplines that are inseparably complementary can at the same time be logically distinguishable, it may be asserted that this particular distinction first became manifest in *Les Ruines des plus beaux monuments de la Grèce* ("The Ruins of the Most Beautiful Monuments of Greece"), written in 1758 by a French architecture student, Julien-David LeRoy. Faced with the problem of discussing Athenian buildings constructed in the time of Vitruvius, he decided to discuss them twice, by treating them separately under two different headings. Before this date, "history" was of architectural importance only as a means of justifying, by reference to classical mythology, the use of certain otherwise irrational elements, such as caryatids. Even Jacques-François Blondel, who in 1750 was probably the first architectural teacher to devote a separate section of his lecture courses to "history," envisaged the subject mainly as an account of the literary references to architecture found in antique manuscripts—an attitude already developed. Renaissance architect Leon Battista Alberti.

Athens: Erechtheum Caryatid porch of the Erechtheum, on the Acropolis at Athens.

The modern concept of architectural history was in fact simply part of a larger trend stimulated by the leading writers of the French Enlightenment, an 18th-century intellectual movement that developed from interrelated conceptions of reason, nature, and man. As a result of discussing constitutional law in terms of its evolution, every branch of knowledge (especially the natural and social sciences) was eventually seen as a historical sequence. In the philosophy of architecture, as in all other kinds of philosophy, the introduction of the historical method not only facilitated the teaching of these subjects but also militated against the elaboration of theoretical speculation. Just as those charged with the responsibility of lecturing on ethics found it very much easier to lecture on the history of ethics, rather than to discuss how a person should or should not act in specific contemporary circumstances, so those who lectured on architectural theory found it easier to recite detailed accounts of what had been done in the past, rather than to recommend practical methods of dealing with current problems.

Moreover, the system of the Paris École des Beaux-Arts (which provided virtually the only organized system of architectural education at the beginning of the 19th century) was radically different from that of the prerevolutionary Académie Royale d'Architecture. Quatremère de Quincy, an Italophile archaeologist who had been trained as a sculptor, united the school of architecture with that of painting and sculpture to form a single organization, so that, although architectural students were ultimately given their own professor of theory, the whole theoretical background of their studies was assimilated to the other two fine arts by lecture courses and textbooks such as Hippolyte Taine's Philosophie de l'art, Charles Blanc's Grammaire des arts du dessin, and Eugène Guillaume's Essais sur la théorie du dessin.

Similarly, whereas before 1750 the uniformity of doctrine (the basic premises of which were ostensibly unchanged since the Renaissance) allowed the professor of architecture to discuss antique and buildings as examples of architectural theory and to ignore medieval buildings completely, the mid-19th-century controversy between "medievalists" and "classicists" (the "Battle of the Styles") and the ensuing faith in Eclecticism turned the studies of architectural history into courses on archaeology.

Thus, the attitudes of those scholars who, during the 19th and early 20th centuries, wished to expound a theory of architecture that was neither a philosophy of art nor a history of architecture tended to become highly personal, if not idiosyncratic. By 1950 most theoretical writings concentrated almost exclusively on visual aspects of architecture, thereby identifying the theory of architecture with what, before 1750, would have been regarded as simply that aspect that Vitruvius called venustas (i.e., "beauty"). This approach did not necessarily invalidate the conclusions reached, but many valuable ideas then put forward as theories of architecture were only partial theories, in which it was taken for granted that theoretical concepts concerning construction and planning were dealt with in other texts.

Distinction Between the Theory of Architecture and the Theory of Art

Before embarking on any discussion as to the nature of the philosophy of architecture, it is essential to distinguish between two mutually exclusive theories that affect the whole course of any such speculation. The first theory regards the philosophy of architecture as the application of a general philosophy of art to a particular type of art. The second, on the contrary, regards the philosophy of architecture as a separate study that, though it may well have many characteristics common to the theories of other arts, is generically distinct.

The first notion (i.e., that there exists a generic theory of art of which the theory of architecture is a specific extension) has been widely held since the mid-century, when the artist and writer Giorgio Vasari published in his *Le vite de' più eccellenti pittori, scultori ed architettori italiani* (*The Lives of the Most Eminent Italian Painters, Sculptors and Architects*) his assertion that painting, sculpture, and architecture are all of common ancestry in that all depend on the ability to draw. This idea became particularly prevalent among English-speaking theorists, since the word design is used to translate both *disegno* ("a drawing") and *concetto* ("a mental plan"). But its main influence on Western thought was due to Italophile Frenchmen, after Louis XIV had been induced to establish in Rome a French Academy modelled on Italian art academies.

As a result of the widespread influence of French culture in the 17th and 18th centuries, the concept of the *beaux arts* (literally "beautiful arts" but usually translated into English as "fine arts") was accepted by Anglo-Saxon theorists as denoting a philosophical entity, to the point where it was generally forgotten that in France itself the architectural profession remained totally aloof from the Académie Royale de Peinture et de Sculpture until they were forced to amalgamate after the French Revolution.

This theory of fine art might not have been so widely adopted but for the development of aesthetics, elaborated after 1750. Thus, when academies of fine art were being established successively in Denmark, Russia, and England on the model of the French Academy in Rome, German philosophers were gradually asserting (1) that it was possible to elaborate a theory of beauty without reference to function (*Zweck*); (2) that any theory of beauty should be applicable to all sensory perceptions, whether visual or auditory; and (3) that the notion of beauty was only one aspect of a much larger concept of life-enhancing sensory stimuli.

The alternative theory (i.e., that a philosophy of architecture is unique and can therefore be evolved only by specific reference to the art of building) will be dealt with below with reference to the traditional triad usually cited in the formula coined, by the English theorist Sir Henry Wotton, in his book *The Elements of Architecture*, namely "commodity, firmness, and delight."

Generally speaking, writers on aesthetics have been noticeably reluctant to use architectural examples in support of speculations as to the nature of their general theories, but references to buildings have been used in most "philosophies of art" ever since the German philosophers Immanuel Kant and G.W.F. Hegel first popularized the philosophical discipline. Kant, in his *Kritik der Urteilskraft*, distinguished between what he termed free beauty (*pulchritudo vaga*) and dependent beauty (*pulchritudo adhaerens*). He classified architecture as dependent beauty, saying that in a thing that is possible only by means of design (*Absicht*)—a building or even an animal—the regularity consisting in symmetry must express the unity of the intuition that accompanies the concept of purpose (*Zweck*), and this regularity belongs to cognition. Nevertheless, he claimed that a flower should be classified as free beauty (where the judgment of taste is "pure") "because hardly anyone but a botanist knows what sort of thing a flower ought to be; and even he, though recognizing in the flower the reproductive organ of the plant, pays no regard to this natural purpose if he is passing judgment on the flower by taste." What Kant's reaction would have been to a modern plastic imitation flower is impossible to guess, but it will readily be perceived (1) why those who, in the 19th century, accepted the notion that beauty in architecture is *pulchritudo adhaerens* felt such antipathy toward "shams," (2) how the distinction between "pure art" and "functional art" (*Zweckkunst*) became confused, and (3) why there arose a tendency to pursue definitions of "pure beauty" or "pure art" without specifically referring to the function and structure of any particular class of beautiful or artistic objects, such as buildings.

This latter tendency was reinforced when the French philosopher Victor Cousin, writing in 1835, classified the history of philosophy under three distinct headings: the true, the beautiful, and the good. The ensuing acceptance of the idea that beauty was to be studied independently of truth and goodness produced a tendency not merely to regard beauty as something added to a building (rather than conceptually inseparable from the truth and goodness of its structure and function) but to regard beauty as limited to visual and emotional qualities.

In the first half of the 20th century, philosophers grew less dogmatic about aesthetics. But its influence on theories of architecture became stronger because of the popular view that sculpture was essentially nonrepresentational. Thus, although the assertion that "aesthetically, architecture is the creation of sculpture big enough to walk about inside" was meaningful in the 20th century, it would have seemed nonsensical to any architectural theorist living before 1900, when sculpture was invariably thought of either as representational or as a carved refinement of load-bearing wood or stone.

Architectural Types

Architecture is created only to fulfill the specifications of an individual or group. Economic law prevents architects from emulating their fellow artists in producing works for which the demand is nonexistent or only potential. So the types of architecture depend upon social formations and may be classified according to the role of the patron in the community. The types that will be discussed here—domestic, religious, governmental, recreational, welfare and educational, and commercial and industrial—represent the simplest classification; a scientific typology of architecture would require a more detailed analysis.

Domestic Architecture

Domestic architecture is produced for the social unit: the individual, family, or clan and their dependents, human and animal. It provides shelter and security for the basic physical functions of life and at times also for commercial, industrial, or agricultural activities that involve the family unit rather than the community. The basic requirements of domestic architecture are simple: a place to sleep, prepare food, eat, and perhaps work; a place that has some light and is protected from the weather. A single room with sturdy walls and roof, a door, a window, and a hearth are the necessities; all else is luxury.

"Vernacular" Architecture

In much of the world today, even where institutions have been in a continuous process of change, dwelling types of ancient or prehistoric origin are in use. In the industrialized United States, for instance, barns are being built according to a design employed in Europe in the 1st millennium BCE. The forces that produce a dynamic evolution of architectural style in communal building are usually inactive in the home and farm. The lives of average people may be unaltered by the most fundamental changes in their institutions. The people can be successively slaves, the subjects of a monarchy, and voting citizens without having the means or the desire to change their customs, techniques, or surroundings. Economic pressure is the major factor that causes average individuals to restrict their demands to a level far below that which the technology of their time is capable of maintaining. Frequently they build new structures with old techniques because experiment and innovation are more costly than repetition. But in wealthy cultures economy permits and customs encourage architecture to provide conveniences such as sanitation, lighting, and heating, as well as separate areas for distinct functions, and these may come to be regarded as necessities. The same causes tend to replace the conservatism of the home with the aspirations of institutional architecture and to emphasize the expressive as well as the utilitarian function.

"Power" Architecture

As wealth and expressive functions increase, a special type of domestic building can be distinguished that may be called power architecture. In almost every civilization the pattern of society gives to a few of its members the power to utilize the resources of the community in the construction of their homes, palaces, villas, gardens, and places of recreation. These few, whose advantages usually arise from economic, religious, or class distinctions, are able to enjoy an infinitevariety of domestic activities connected with the mores of their position. These can include even communal functions: the palace of the Flavian emperors in ancient Rome incorporated the activities of the state and the judicial system; the palace of Versailles, a whole city in itself, provided the necessities and luxuries of life for several thousand persons of all classes and was the centre of government for the empire of Louis XIV. Power architecture may have a complex expressive function, too, since the symbolizing of power by elegance or display is a responsibility or a necessity (and often a fault) of the powerful. Since this function usually is sought not so much to delight the patrons as to demonstrate their social position to others, power architecture becomes communal as well as domestic. In democracies such as ancient Greece and in the modern Western world, this show of power may have been more reserved, but it is still distinguishable.

Versailles, Palace of Palace of Versailles, France

Group Housing

A third type of domestic architecture accommodates the group rather than the unit and is therefore public as well as private. It is familiar through the widespread development of mass housing in the modern world, in which individuals or families find living space either in multiple dwellings or in single units produced in quantity. Group housing is produced by many kinds of cultures: by communal states to equalize living standards, by tyrants to assure a docile labour force, and by feudal or caste systems to bring together members of a class. The apartment housewas developed independently by the imperial Romans of antiquity to suit urban conditions and by the American Indians to suit agricultural conditions. Group architecture may be power architecture as well, particularly when land values are too high to permit even the wealthy to build privately, as in the 17th-century Place des Vosges in Paris, where aristocratic mansions were designed uniformly around a square, or in the 18th-century flats in English towns and spas. Although most domestic architecture of the 20th century employed the style and techniques of the past, the exceptions are more numerous and more important for the development of architecture than ever before. This is because the distribution of wealth and power is widespread in parts of the world where architecture is vital and because the modern state has assumed responsibility for much high-quality housing.

Interior Design

Interior design, planning and design of man-made spaces, a part of environmental design and closely related to architecture. Although the desire to create a pleasant environment is as old as civilization itself, the field of interior design is relatively new.

Since at least the middle of the 20th century, the term interior decorator has been so loosely applied as to be nearly meaningless, with the result that other, more descriptive terms have come into use. The term interior design indicates a broader area of activity and at the same time suggests its status as a serious profession. In some European countries, where the profession is well established, it is known as interior architecture. Individuals who are concerned with the many elements that shape man-made environments have come to refer to the total field as environmental design.

Principles of Interior Design

It is important to emphasize that interior design is a specialized branch of architecture or environmental design; it is equally important to keep in mind that no specialized branch in any field would be very meaningful if practiced out of context. The best buildings and the best interiors are those in which there is no obvious disparity between the many elements that make up the totality. Among these elements are the structural aspects of a building, the site planning, the landscaping, the furniture, and the architectural graphics (signs), as well as the interior details. Indeed, there are many examples of distinguished buildings and interiors that were created and coordinated by one guiding hand.

Because of the technological complexity of contemporary planning and building, it is no longer possible for a single architect or designer to be an expert in all the many aspects that make up a modern building. It is essential, however, that the many specialists who make up a team be able to communicate with each other and have sufficient basic knowledge to carry out their common goals. While the architect usually concerns himself with the overall design of buildings, the interior designer is concerned with the more intimately scaled aspects of design, the specific aesthetic, functional, and psychological questions involved, and the individual character of spaces.

Although interior design is still a developing profession without a clear definition of its limits, the field can be thought of in terms of two basic categories: residential and nonresidential. The latter is often called contract design because of the manner in which the designer receives his compensation (*i.e.*, a contractual fee arrangement), in contrast to the commission or percentage arrangement prevalent among residential interior decorators. Although the volume of business activity in the field of residential interiors continues to grow, there seems to be less need and less challenge for the professional designer, with the result that more and more of the qualified professionals are involved in nonresidential work.

The field of interior design already has a number of specialized areas. One of the newer areas is "space planning"—*i.e.*, the analysis of space needs, allocation of space, and the interrelation of functions within business firms. In addition to these preliminary considerations, such design firms are usually specialists in office design.

Many design firms have become specialized in such fields as the design of hotels, stores, industrial parks,or shopping centres. Others work primarily on large college or school projects, and still others may be specialists in the design of hospitals, clinics, and nursing homes. Design firms active in nonresidential work range from small groups of associates to organizations comprised of 50 to 100 employees. Most of the larger firms include architects, industrial designers, and graphic designers. In contrast, interior designers who undertake residential commissions are likely to work as individuals or possibly with two or three assistants. The size of the firms involved in nonresidential design is a clear indication of the relative complexity of the large commissions. In addition to being less complex, residential design is a different type of activity. The residential interior is usually a highly personal statement for both the owner and the designer, each of whom is involved with all aspects of the design; it is unlikely that a client who wished to engage the services of an interior designer for his home would be happy with an organized systems approach.

Most large architectural firms have established their own interior-design departments, and smaller firms have at least one specialist in the field. There are no precise boundaries to the profession of interior design nor, in fact, to any of the design professions. Furniture design, for example, is carried out by industrial designers and furniture designers as well as by architects and interior designers. As a rule, furniture designed for mass production is designed by industrial designers or furniture designers; the interior designer or architect usually designs those special pieces that are not readily available on the market or that must meet specific needs for a particular job. Those needs may be functional or aesthetic, and often a special chair or desk designed for a specific job will turn out to be so successful that the manufacturer will put such pieces into his regular line. The same basic situation holds generally true in the design of fabrics, lighting devices, floor covering, and all home-furnishing products. All design activities are basically similar, even though the training and education in the different design fields varies in emphasis. A talented and well-trained designer can easily move from one specialized area to another with little difficulty.

In the discussion of the general aspects of design, it is important to note that there is an important distinction between art and design. A designer is basically concerned with the solution of problems (be they functional, aesthetic, or psychological) that are presented to him. The artist is more concerned with emotive or expressive ideas and with the solution of problems he himself poses. A truly great or beautiful interior can indeed be called a work of art, but some would prefer to call such an interior a "great design."

Aesthetic Components of Design

A general definition of beauty and aesthetic excellence would be difficult, but fortunately there are a number of generally accepted principles that can be used to achieve an understanding of the aesthetic considerations in design. One must note, however, that such understanding requires exposure and learning; an appreciation of any form of art needs such a background.

A thorough appreciation of design must go beyond the first impression. The first impression of the interior of a Gothic cathedral might be that it is somewhat dark or gloomy, but, by the time the visitor senses its majestic proportions, notices its beautiful stained glass windows and the effect of light, and begins to understand the superb structural system that permitted builders of cathedrals to achieve their lofty goals, he can truly begin to appreciate the overall aesthetic qualities.

One of the key considerations in any design must be the question of whether a design "works" or functions for its purpose. If a theatre has poor sight lines, poor acoustics, and insufficient means of entry and egress, it obviously does not work for its purpose, no matter how beautifully it might be decorated. Such a design could be considered good only if it were thought of abstractly as a kind of walk-in sculpture. In some cases the building is meant to be sculpture rather than architecture. The Statue of Liberty, for instance, is primarily intended as a monument, despite the fact that it contains rather tortured interior spaces.

To use function as the only aesthetic criterion would be limiting, but it certainly is a valid consideration to be kept in mind. Designers are often tempted to overdesign or "style" an object or interior rather than design it. Some of the most beautiful objects of the 20th century are beautiful because they were the result of purely functional considerations. It is conceivable that future art historians will consider a modern jet plane the crowning artistic achievement of the middle of this century, rather than any building, interior, or conscious art form.

The aesthetic response to an interior and its furnishings must take into consideration the social and economic conditions as well as the materials and technology of the time. The elegant or ornate interiors that are usually associated with the 18th and 19th centuries were appropriate to the social and economic conditions of the nobility or the wealthy bourgeois who were the original occupants. The chairs were designed for formal living, and the elaborately carved furnishings were designed to be cared for by many servants. Such an interior is alien to the 20th-century way of life and would be totally inappropriate for a contemporary middle class family. It would also be inappropriate to use modern materials and processes to imitate earlier materials and processes. Many manufacturers try desperately to make plastic look like wood, stone, or just about anything but plastic. All aesthetic criteria have something to do with honesty. Some aestheticians have compared beauty to truth, and there can be little doubt that honestly expressed functions and honestly expressed materials and manufacturing processes are far more beautiful than fakery and imitation.

All interiors, by definition, occur inside buildings and therefore have a very real relation to these buildings. The best interiors today, as well as in the past, are those that relate well in character and appropriateness to the particular building. The furnishings designed and scaled for spacious country homes or palaces would obviously be out of place in a small urban apartment or suburban home. A strong and unusual piece of architecture such as New York City's Trans World Airlines terminal (at John F. Kennedy International Airport) could not be properly furnished with standard commercial furniture and products. The building, as well as the interiors, was conceived as a total design by the Finnish-born architect Eero Saarinen. Whether the observer agrees with the architect's concept or not, he clearly senses the strong interrelationship between the exterior and the interior—and therefore the aesthetic unity and success. Another successful interior and building is the Ford Foundation headquarters in New York City, the work of architects Kevin Roche and John Dinkeloo, with interiors by Warren Platner. The design is notable for its handsome spaces opening out toward an enclosed garden space. This obviously would not have been possible or appropriate if the view from the offices had been unattractive.

The interiors within indifferent or unattractive buildings must strive to make up for the lack of design qualities in the structures. Thus, it is sometimes necessary to ignore the ugliness of the building and create an inward-looking beauty if no architectural character exists.

The most difficult aesthetic consideration is the problem of appropriateness. The appropriate atmosphere or character of an interior must take all the foregoing points into consideration. The architectural character of the TWA terminal would make it inappropriate for use as an office building. The appropriateness of individual, more intimate, and small-scaled interiors is more subtle. The interior design of a discotheque would hardly be appropriate for a research library, and a college classroom would hardly provide the desired atmosphere for a kindergarten. Many of these responses and relationships are complex and have psychological as well as aesthetic factors.

Elements of Design

Of all the component elements that together form a completed interior, the single most important element is space. Spaces can be exhilarating or depressing, cheerful or serene, all depending upon the use the designer has made of the various elements that form the whole. Space is, in modern times, a costly commodity. The beautiful space of the Gothic cathedral achieved its success through generous proportions and lofty heights. Due to the vast increase in construction costs in contemporary structures, spaces tend to be smaller and less generous; more skill on the part of the designer is required to give such limited spaces a particular atmosphere or character. On the other hand, sheer volume of space is not sufficient. There is hardly a larger space than the interior of the Vehicle Assembly Building at the John F. Kennedy Space Center in Florida, yet the aesthetic impact of that immense interior is negligible. A space need not be large and monumental to be aesthetically successful. The handling of mass and form even within a small structure can become exciting and beautiful. Frank Lloyd Wright was masterful in creating beautiful spatial sequences within residential-scale buildings. The Ford Foundation building is a relatively small structure among the huge buildings of New York City, yet the experience of that space is real and pleasurable.

Space can be thought of as the raw material which must be molded and shaped with the designers' tools of colour, texture, light, and scale. The interrelationship of design elements can be clarified by visualizing the result if the interior of St. Peter's in Rome were painted in garish colours or painted all black or sprayed with a foamy texture covering all surfaces or flooded with enormously intense floodlight that eliminated all play of dark and light. Obviously, any of these modifications would totally destroy the beauty and success of that space.

Colour is the quality of light reflected from an object to the human eye. When light falls upon an object, some of it is absorbed, and that which is not absorbed is reflected, and the apparent colour of an object depends upon the wavelength of the light that it reflects. The scientific attributes of colour and light in interior design are, however, less important than the skillful combination of colour values, hues, tones, shades, and above all textures. Although there can be no strict rules about colours and textures, it is well to remember the famous statement of the modern architect Mies van der Rohe that "less is more." His Crown Hall at Illinois Institute of Technology in Chicago, built in 1956, is elegant, understated, subtle, and is notable for its careful handling of textures and materials. To accept "less is more" as the sole guideline to design, however, would be a serious fallacy. Space, which is the essence of a meaningful interior, would be dull indeed if it were never varied—if there were no intimate spaces with low ceilings, in contrast to large spaces of greater height, and if spaces did not interrelate to provide the user with a sequential experience of moving from one to another. Monotony would also result if all interiors in a given building were of the same colour, material, and textural quality. Man needs variety and change.

The manipulation of space is a matter of both aesthetic and functional consideration. A small entrance vestibule in a building is needed to keep out wind and cold or heat and rain, yet it is equally important in providing a visual transition from outdoors to the interior of the building. The sheltered sleeping alcoves in early cave dwellings served not only to express man's desire for smaller and more intimate spaces for personal use but gave protection from draft or cold.

Much in our man-made structures is built of natural materials, and it must be remembered that these materials have natural colours and textures that usually are superior to anything man can create artificially. Competent designers are very much aware of the innate qualities and textures of all materials, especially natural ones. For instance, a sensitive designer would choose a simple oil finish on wood to bring out the beauty and quality of the grain rather than use the once-fashionable high-gloss finish that tended to obscure and change the texture. Textures are important not only for their appearance but also for their sense of touch, and for their effect on light absorption or reflection. Abrasive surfaces or very rough plaster would obviously be unpleasant to the touch and possibly dangerous in an interior, depending upon the use the interior is intended for. Textures can evoke feelings of elegance (such as silks) or informality (such as rough, tweedy materials).

Light, both natural and artificial, is one of the most important design elements, but unless surfaces are appropriate in colour and texture, the control and effect of light will be lost. The beautiful quality of space in a Gothic cathedral is very much related to the handling of light. The source of daylight, high overhead or filtered through stained glass, creates exciting patterns of light and shade and a variety of intensities and pools of light. This same principle can be used in all interior spaces, and contemporary interiors often have skylights or high windows to provide variety and changing patterns of light. Artificial lighting is equally important, and, again, the same considerations of highlights, good overall illumination, and variety are important.

Concepts of Design

The scale and proportion of any interior must always relate to the architecture within which the interior exists, but the other important factor in considering the scale of man's environment is the human body. Throughout the ages, designers and architects have attempted to establish ideal proportions. The most famous of all axioms about proportion was the golden section, established by the ancient Greeks. According to this axiom, a line should be divided into two unequal parts, of which the first is to the second as the second is to the whole. Leonardo da Vinci developed a figure for the ideal man based on man's navel as the centre of a circle enclosing man with outstretched arms. The French architect Le Corbusier developed a theory of proportion called Modulor, also based on a study of human proportions. Yet, at best, these rules are merely guidelines. They can never substitute for the eye and judgment of the designer, and it is reasonable to predict that attempts to make the all-powerful computer a substitute for the designer's sensitivity are also bound to be far from perfect.

It was stated earlier that the need for a changing scale and spatial relationship in the environment seems a natural one, almost a physiological as well as a psychological one. Perhaps the need for "personal" environment and scale can best be understood by considering some extreme examples. To a person flying at 30,000 feet in an airplane, the scale of anything seen on the ground appears so small that he loses touch with the reality of objects. People who fear heights are rarely bothered by the view out of an airplane because the distance to the objects on the ground has transcended normal perceptions of scale. In a similar manner, a person's reaction to the scale of a small house is quite different from his reaction to a large high-rise building. Details of pattern, texture, and material are accepted and expected in the small structure since they are in a meaningful scale with respect to man. By the same token, the sculptural ornaments on the tops of early skyscrapers seem absurd today.

Almost all principles of design for interiors can be comprehended with clear analytic understanding and common sense, without regard to dogmatic rules. If a beautiful 18th-century breakfront (which might be more than eight feet tall) is placed in an apartment with a ceiling height just an inch higher than the piece of furniture, it would obviously look out of scale. If a space is planned so that all the heavy and massive pieces of furniture are pushed toward one end of the room, with nothing on the other side, the room would obviously look out of balance. Yet balance and symmetry applied as inviolate design principles would result in very formal, very traditional, and somewhat dull interiors. Careful symmetry was a generally accepted rule during the Renaissance, and in any classic building one can be sure to find a carefully balanced and symmetrical facade, just as most formal and classic interiors have rigidly balanced plans. It is now recognized that balance can also be based on asymmetry. Both architecture and interior design in the 20th century have consciously broken with the many rules handed down from past eras. It is more important for a building or space to be expressive of its purpose. At one time, it was traditional for a theatre, opera house, or concert hall to embody certain forms and shapes without any real consideration of sight lines, seating distance from the stage, or acoustics. On the other hand, the Berlin Philharmonic Concert Hall works beautifully as a concert hall and expresses its purpose and function clearly in an exciting and dynamic way.

Balance and symmetry, colour, pattern, and repetition used to be a matter of adherence to a tradition. Until fairly recently, many interiors were painted in dark colours, often ignoring the fact that light reflection was adversely affected and that no real contrast or sparkling accent was achieved. In many contemporary rooms, however, most surfaces are kept in neutral or light colours, possibly with one wall accented in a strong colour or texture. An interior with uniform overhead lighting might be an efficient work space but would lack the character that can be achieved by providing some accent lights in small areas.

The designer's concern for honesty of materials and textures has brought about changing attitudes toward some of the conventional practices of interior decoration, such as the use of strongly patterned wallpapers and flowered prints. Any interior that has too many different patterns, too many textures, and too many repetitive features of any kind will appear overpowering, overly busy, overdesigned, and confusing. A designer often attempts to have a dominant theme or idea, be it colour, form, texture, or some rhythmic pattern. It must be noted also that design is influenced by changing attitudes and fashions. The movements in art and architecture of the 1950s and 1960s have influenced interior design in the direction of an emphasis on pure form, the absence of superfluous decoration, and expressiveness of materials. Recently, however, a kind of countermovement in the field of painting and sculpture has been influential. For instance, the use of large-scale graphic elements (supergraphics) in interiors has become popular and accepted, in spite of the fact that its very idea often consciously denies or destroys the visual clarity of existing architectural design features. Some of the leading designers in the United States and in several European countries have also become very interested in large patterns, rhythmic geometries, and decorative surfaces, and this may point toward a new trend.

Most interiors consist of a series of interrelated spaces. It is important that the various spaces be designed in a sequential relationship to each other, not only in terms of planning but also in terms of the visual effect. A successful interior should be cohesive within each area and cohesive as a totality. It must above all relate to the building and to the architectural concept. A good example is the previously mentioned TWA terminal by Eero Saarinen. In spite of the extremely complex sculptural forms used, there is a sequence and clearly balanced rhythm that not only unifies the total composition but clearly relates it to the total architecture.

Supergraphic interior emphasizing decorative rather than architectural design: Hear-Hear Record Shop, San Francisco, designed by Daniel Solomon, graphics designed by Barbara Stauffacher.

The best examples of design are those in which no visible difference exists between the interior and the exterior, between the building and its site, and between the many parts or spaces to each other and the total building. An example is the house of the American architect Philip Johnson in New Canaan, Connecticut. Johnson's home and its setting appear effortlessly united, with individual parts subordinated to the success of the whole.

Interrelation of interior and exterior space. Harmony of landscape, architecture, and interior design: (top) exterior and (bottom) interior of the Glass House, New Canaan, Connecticut.

Design Relationships

The real and conscious relationship between art, architecture, and design is of long standing. Though mural painting was largely neglected in the mid-20th century, in the past great murals have been the planned focal points of interiors and have in a way determined the architecture. Similarly, sculpture or sculptural forms, as fixed and permanent spects of buildings, can be the most important design features if planned that way by the architect together with the interior designer and artist. Perhaps the best design is one in which there is no visible difference between architecture and interior and in which even the artwork is incorporated as an integral part of the total.

The design relationship of interiors to architecture can be clarified by citing an extreme example: the stage set. A set for a theatrical production is a form of interior design but, unlike all other aspects of interior design, it attempts to create its own world and atmosphere concerned only with the play and not at all related to the world or even reality. The creation of a world of make-believe is precisely the function of a stage, but in real life it is impossible to divorce a particular interior from everything else around it. Sometimes a designer may attempt to create a "theatrical" interior, but the point being made strongly and unequivocally here is that every interior must relate to the architecture and to the nearby environment.

A ramp functioning as the focal element of an interior: the former V.C.
Morris Shop, San Francisco, designed by Frank Lloyd Wright.

Design relationships of individual works of art (paintings, prints, or sculptures) to interiors are most significant in terms of scale and placement, rather than in terms of subject matter, colour, or style. A very old painting, if it is good, will look well within a contemporary interior; a very modern piece of sculpture can be beautiful within an interior furnished with some beautiful traditional pieces. Any work of art, if successful within itself, is "correct" with any interior if properly placed or selected to work with the total space. Certainly there is no need to match colours of paintings to interiors or to select subject matter in works of art that reflect a particular theme, such as food for dining rooms or hunting scenes for the den.

Interiors as they relate to landscape or cityscape are sometimes misunderstood by architects. A crass but typical example is the ubiquitous picture window in suburban housing tracts. Often the only view from the window is the picture window of the neighbouring house. When the view is a beautiful one, it should be possible to plan the interior with the furniture plan and orientation such that seating arrangements can take advantage of the view and yet work for other functions, such as relation to a fireplace or a conversation group, as well.

In many areas of interior design the field of graphics is taking on increasing importance. In every public or institutional building, signs, directories, and room identifications play an important visual part. Good architectural graphics have been stressed only in recent years, as a result of the increasing size and complexity of structures. Buildings such as airports depend upon clear and handsome graphics to make the spaces work and to make them aesthetically cohesive. A related aspect of graphics is the printed matter that is part of certain interior functions. Interior designers must be concerned with the design of menus, wine lists, napkins, and matchbooks in a well-designed restaurant. Designers dealing with stores or shops are concerned with the graphics of shopping bags, signs, and posters. Often the interior designer is the actual graphic designer, or he works together with the graphic designer, just as the architect works with the interior designer or landscape architect.

Modes of Composition

It must be emphasized that there are many different moods, or modes of composition, that are possible in interior design. The recognition of this fact makes it difficult to apply valid critical criteria to these modes, since many of them are intensely personal. What may appear to be picturesque to one person might be ugly or cluttered to another. Each person brings to interior design his own cultural mores and his own prejudices, and in many ways he is psychologically conditioned and influenced to accept certain things and to reject others. In discussing various modes of composition, one must therefore take into consideration the occupants and their backgrounds, the locale and site, and then try to apply the most basic design principles as general guidelines.

Formal and informal compositions are relatively easily defined and classified; in fact, this distinction is useful throughout the history of furniture and interiors. Formal styles are usually associated with life at court or furnishings for the palatial homes of nobles or a moneyed elite. The informal periods usually are associated with rural living or the simpler pieces of furniture made by the local craftsmen in rural areas, where they plied their trade with limited tools, using local woods. Formal furniture, as a rule, leads to formal interior compositions. Balance and symmetry certainly tend to lead to formal compositions. Formality is not associated with any particular period. In fact, a very famous contemporary chair, the Barcelona chair by Mies van der Rohe, is an extremely formal and elegant piece. It would seem wrong to use that chair in a casual catercorner room arrangement.

Setting strongly influences the character of a space. By its very definition, a rustic setting would be rural and informal and would seem wrong and incongruous in a formal townhouse or city apartment. Since most business and public interiors are located in urban centres, any attempt to make such interiors look rustic or homey would be an aesthetic paradox. By the same token, it would appear equally incongruous to design a restaurant located in an old mill or barn in New England in a formal and urban character with elegant furnishings, whether they were contemporary or antiques of a formal nature.

Certain modes of composition are determined by the function of the spaces as much as by the location and by the architecture. For example, a cozy or homey interior is normally associated with residential interiors or similarly intimate interiors, such as restaurants that may wish to appear "cozy." Some interiors, such as discotheques, require excitement and other interiors, such as funeral parlors, require serenity or dignity. One expects certain modes of composition for certain functions, but one's expectations are subject to many external influences, such as personal background, location, psychological associations, and changing fashions. For instance, the typical bank interior until about 1950 was expected to be solid, dignified, awe-inspiring, formal, and above all confidence inspiring. Contemporary design for business and industry has become accepted by all, and the early 1950s saw the logical extension of these firmly established design principles into the area of bank design. One of the first radical departures of traditional design for banking spaces was the Manufacturers Trust Company Manhattan office designed by Skidmore, Owings and Merrill in the early 1950s. It was the first widely published "glass" bank, and it set a trend that has become the new mode of composition for banks.

Fashion or design trends influence one's reactions to many kinds of designs. The term clutter is usually associated with Victorian design of the 19th century. Under the usual definition of the term clutter, one thinks of home interiors with collections of accessories and with an overabundance of knickknacks—the typical Victorian home. In the mid-1960s a new approach to office design, reflecting the "cluttered" approach, was developed. This office appears disorganized at first glance. Actually, the system is very efficient and for that reason is deemed acceptable, even if the visual impact tends to be chaotic. Traditionally, office and business interiors were pristine, orderly, and very organized, and the idea of a cluttered appearance would have been anathema to designers.

The most difficult mode of composition for objective analysis is one that some people call exotic. The chances are that all exotic interiors are highly personal statements and cannot be rationally understood in theoretical design terms. To begin with, what may appear exotic to the average American could be very ordinary or even homey to another culture. Japanese or oriental design in general serves as an example. A Japanese style interior is extremely subtle, serene, and understated, yet to the uninitiated such an interior will appear exotic. Undoubtedly that same phenomenon holds true in reverse. Oriental people have often been impressed with Western-style design and have adopted it presumably because to them it appeared exotic. The increased mobility of the middle classes of many nations today has made foreign travel possible for more and more people, thereby tending to soften some of the very strong regional differences in design. The modes of composition are still discernible nationally or certainly by major geographic and ethnic divisions, but they tend to be less distinct. Many subtle differences exist within the same country, some of which are based on varying socioeconomic backgrounds, much in the manner of the traditional difference between formal styles (at court and in homes of nobility) and informal modes of composition for the country people and middle classes. The labels that one applies to these modes of composition are often only descriptive. They must not be confused with objective evaluation of design values. An interior that is by the creator's definition exotic or picturesque may or may not be a well-done exotic design.

Symbolism and Style

There are many historic examples of symbolism in design, but often the symbolism is not a conscious statement so much as a more subtle reflection of style. Religious buildings, especially churches, have until recently been consistently traditional expressions of style or symbolism. The church and church architecture flourished during the Middle Ages, and the style of church architecture that became the dominant symbol was the Gothic style. Until the recent past, churches were still designed, as a matter of course, in Gothic style. It is interesting to note that a "Gothic" church designed and built in 1820 can be clearly identified as such, and a "Gothic" church from the year 1920 has the imprint of that year as obviously as the date on its cornerstone. There has been a similar symbolic or stylistic tradition in the design of public or governmental buildings. Both interiors and exteriors of city halls, court buildings, and major government structures were usually in the "classical" style, symbolizing authority, power, and stability, based on our long historic association of these concepts with Greco-Roman antiquity and Renaissance thought.

Another form of symbolism in interior design has been the creation of interiors around specific themes or concepts. Among the earliest examples is the Egyptian tomb. The interior design and decoration depicted the life of the king or special events from his life, and the total interior was intended as a kind of magic to assure the occupant's journey into life after death and guarantee his happiness there. Another example of a symbolic interior created for a specific purpose is the Roman hunting lodge, Piazza Amerina, in Sicily, which has splendid murals and floors depicting animals and hunting. A more recent example of a similarly symbolic interior on the same subject is Theodore Roosevelt's home at Oyster Bay on Long Island, built in 1880. It is full of hunting trophies and mementos symbolizing his personal interests and his personality.

The styles that developed in interiors and in interior furnishings were always symbolic of the social structure of the society that created them. It is easy, for instance, to look at the graceful, feminine lines of a Louis XV chair, delicately curved and luxuriously upholstered, and to see it as a symbolic expression of the superficialities of court life. One can also look at some of the crudely fashioned early American furniture and see in one's mind the life of the settler who fashioned it. Life was harsh, time was precious, and articles of furniture were confined to essentials. The need for economical use of space was symbolized by dual-purpose, functional pieces such as dough boxes that served as tables and tables that turned into chairs and had storage compartments for the family Bible as well.

As functional and efficiency-oriented as business and office design is today, it is full of unwritten rules relating to symbolism. The design of an office reflects the status of the occupant. Top executives are located in the largest corner offices with the best views of the city and invariably are on the top floors of the corporate headquarters. The size of desks is a symbolic indication of the executive's importance in the hierarchy of the firm. The very top officers may, however, do away with desks altogether and have offices resembling living rooms—to symbolize the fact that they are beyond routine paperwork and above the need for standard office furnishings. The fashions (or styles) of design vary and develop even within a brief period of 10 or 20 years. Thus, another symbol—carpeting—has become somewhat outdated. Until recently, top executives expected wall-to-wall carpeting in their offices. Today such offices may have wood or other natural floors, perhaps with beautiful area rugs. The very idea of a private office is, of course, the most important symbol in a status-conscious business community. Designers have found, however, that the need for communication between executive and staff, including visual contact, often makes private offices less than efficient.

Symbolism in residential interior design occurs on many levels but again tends to be influenced by changing styles. When television first became available, the home screen became a symbol of prosperity and at the same time became the focal point of residential interiors. By the 1970s a television set had become a standard possession and was no longer a compositional emphasis; in fact, it was often concealed or casually incorporated into the total design.

A homeowner is likely to be very conscious of the image his house or apartment conveys. Traditional furniture, for instance, is still associated with elegance in the minds of many laymen, a situation that can lead to the acquisition of poor reproductions or meaningless imitations of non-existent styles. To most people a real fire in a fireplace is a delightful physical and visual experience that often has nostalgic associations. Since they are no longer needed to heat houses, fireplaces in the 20th century increasingly have become a luxury and thereby a symbol of substance to many people. These circumstances have often resulted in imitation fireplaces of the worst possible design, with simulated fires.

From the designer's point of view, design symbolism in public spaces is valid at times but can and should be used in contemporary terms rather than as stylistic imitation of past eras. An example of the success of such design can be seen in the new Boston City Hall, built in 1968, which symbolizes government, authority, and dignity in totally original and contemporary terms. There is little valid reason to consciously introduce symbolism into residential interiors, unless it is the kind of cultural symbolism exemplified in Japanese interiors, such as that of the Zen tea house (*cha-shitsu*), where certain design features reflect a way of life and have ceremonial meanings.

Sustainable Architecture

Sustainable architecture refers to the practice of designing buildings which create living environments that work to minimize the human use of resources. This is reflected both in a building's construction materials and methods and in its use of resources, such as in heating, cooling, power, water, and wastewater treatment.

The operating concept is that structures so designed "sustain" their users by providing healthy environments, improving the quality of life, and avoiding the production of waste, to preserve the long-term survivability of the human species.

Hunter and Amory Lovins of the Rocky Mountain Institute say the purpose of sustainable architecture is to "meet the needs of the present without compromising the ability of future generations to meet their own needs."

The term, however, is a broad one, and is used to describe a wide variety of aspects of building design and use. For some, it applies to designing buildings that produce as much energy as they consume. Another interpretation calls for a consciousness of the spiritual significance of a building's design, construction, and siting. Also, some maintain that the buildings must foster the spiritual and physical wellbeing of their users.

One school of thought maintains that, in its highest form, sustainable architecture replicates a stable ecosystem . According to noted ecological engineer David Del Porto, a building designed for sustainability is a balanced system where there are no wastes, because the outputs of one process become the inputs of another. Energy, matter, and information are cascaded through connected processes in cyclical pathways, which by virtue of their efficiency and interdependence yield the matrix elements of environmental and economic security, high quality of life, and no waste. The constant input of the sun replenishes any energy lost in the process.

Sustainability, as it relates to resources, became a widely used term with Lester Brown's book, *Building a Sustainable Society,* and with the publishing of the International Union on the Conservation of Nature's "World Conservation Strategy" in 1980.

Sustainability then came to describe a state whereby natural renewable resources are used in a manner that does not eliminate or degrade them or otherwise diminish their renewable usefulness for future generations, while maintaining effectively constant or non-declining stocks of natural resources such as soil , groundwater , and biomass (World Resources Institute).

Before "sustainable architecture," the term "solar architecture" was used to express the architectural approach to reducing the consumption of natural resources and fuels by capturing solar energy. This evolved into the current and broader concept of sustainable architecture, which expands the scope of issues involved to include water use, climate control, food production, air purification, solid waste reclamation , wastewater treatment, and overall energy efficiency . It also encompasses building materials, emphasizing the use of local materials, renewable resources and recycled materials, and the mental and physical comfort of the building's inhabitants. In addition, sustainable architecture calls for the siting and design of a building to harmonize with its surroundings.

The United Nations lists the following five principles of sustainable architecture:

- Healthful interior environment. All possible measures are to be taken to ensure that materials and building systems do not emit toxic substances and gasses into the interior atmosphere . Additional measures are to be taken to clean and revitalize interior air with filtration and plantings.

- Resource efficiency. All possible measures are to be taken to ensure that the building's use of energy and other resources is minimal. Cooling, heating, and lighting systems are to use methods and products that conserve or eliminate energy use. Water use and the production of wastewater are minimized.

- Ecologically benign materials. All possible measures are to be taken to use building materials and products that minimize destruction of the global environment . Wood is to be selected based on non-destructive forestry practices. Other materials and products are to be considered based on the toxic waste output of production. Many practitioners cite an additional criterion: that the long-term environmental and societal costs to produce the building's materials must be considered and prove in keeping with sustainability goals.

- Environmental form. All possible measures are to be taken to relate the form and plan of the design to the site, the region, and the climate. Measures are to be taken to "heal" and augment the ecology of the site. Accommodations are to be made for recycling and energy efficiency. Measures are to be taken to relate the form of building to a harmonious relationship between the inhabitants and nature.

- Good design. All possible measures are to be taken to achieve an efficient, long-lasting, and elegant relationship of area use, circulation, building form, mechanical systems, and construction technology. Symbolic relationships with appropriate history, the Earth, and spiritual principles are to be searched for and expressed. Finished buildings shall be well built, easy to use, and beautiful.

The NMB Bank headquarters in Amsterdam, the Netherlands, is an example of sustainable architecture. Constructed in 1978, this approximately 150,000-ft² (45,500-m²) complex is a meandering S-curve of 10 buildings, each offering different orientations and views of gardens. Constructed of natural and low-polluting materials, the buildings feature organic design lines, indoor and outdoor gardens, passive solar elements, heat recovery, water features, and natural lighting and ventilation. Built for an estimated 5% more than a conventional office building, the NMB building's operating costs are only 30% of those of a conventional building. Another example is the Solar Living Center in Hopland, California, which employs both passive and photovoltaic solar elements, as well as ecological wastewater systems. The rice straw bale and cement building is constructed around a solar calendar.

Sustainable Architecture as a Movement

Some maintain that sustainability, as it relates to architecture, refers to a process and an attitude or viewpoint. Sustainability is "a process of responsible consumption, wherein waste is minimized, and buildings interact in balanced ways with natural environments and cycles, balancing the desires and activities of humankind within the integrity and carrying capacity of nature, and achieving a stable, long-term relationship within the limits of their local and global environment."

However, sustainable architecture does not necessarily mean a reduction in material comfort. Sustainability represents a transition from a period of degradation of the natural environment (as represented by the industrial revolution and its associated unplanned and wasteful patterns of growth) to a more humane and natural environment. It is doing more with less.

Proponents of sustainable architecture occasionally debate the broader applications of the term. Some say that sustainable buildings should generate more energy over time (in the form of power, etc.) than was required to construct, fabricate their materials, operate, and maintain them. This is also referred to as "regenerative architecture," which John Tillman Lyle sums up in his book, *Regenerative Design for Sustainable Development*, as "living on the interest yielded by natural resources rather than the capital." Others simply see it as an approach to making buildings less consumptive of natural resources.

Spiritual Aspects of Sustainable Architecture

A spiritual viewpoint is that sustainable architecture is "stewardship," a recognition and celebration of the human environment as a vital part of the larger universe and of humankind's role as caretakers of the earth. Viewed in this way, resources are regarded as sacred. Another perspective is that the creation of a building in the likeness of a living system is somewhat religious, as a divine entity creates a living order.

Although the term communicates slightly different meanings to various audiences, it nevertheless serves as a consciousness-raising focus for creating greater concern for the built environment and its long-term viability. Rather than representing a return to subsistence living, buildings designed for sustainability aim to improve the quality and standards of living. Sustainable architecture recognizes people as temporary stewards of their environments, working toward a respect for natural systems and a higher quality of life.

Sustainable Energy Use

K2 sustainable apartments in Windsor, Victoria, Australia by DesignInc features passive solar design, recycled and sustainable materials, photovoltaic cells, wastewater treatment, rainwater collection and solar hot water.

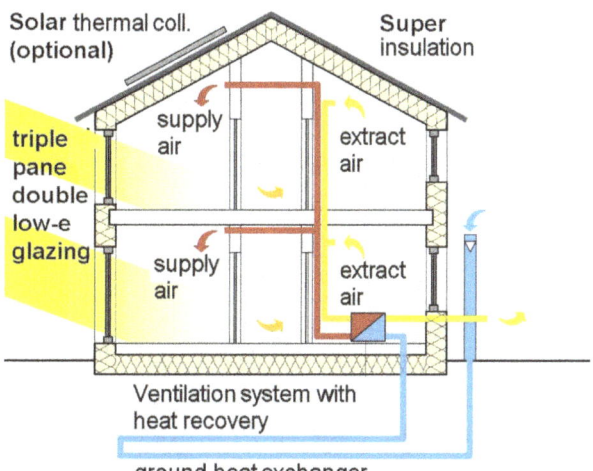

The passivhaus standard combines a variety of techniques and technologies to achieve ultra-low energy use.

Following its destruction by a tornado, the town of Greensburg, Kansas (United States) elected to rebuild to highly stringent LEED Platinum environmental standards. Shown is the town's new art center, which integrates its own solar panels and wind generators for energy self-sufficiency.

Energy efficiency over the entire life cycle of a building is the most important goal of sustainable architecture. Architects use many different passive and active techniques to reduce the energy needs of buildings and increase their ability to capture or generate their own energy. One of the keys to exploit local environmental resources and influence energy-related factors such as daylight, solar heat gains and ventilation is the use of site analysis.

Heating, Ventilation and Cooling System Efficiency

Numerous passive architectural strategies have been developed over time. Examples of such strategies include the arrangement of rooms or the sizing and orientation of windows in a building, and the orientation of facades and streets or the ratio between building heights and street widths for urban planning.

An important and cost-effective element of an efficient heating, ventilating, and air conditioning (HVAC) system is a well-insulated building. A more efficient building requires less heat generating or dissipating power, but may require more ventilation capacity to expel polluted indoor air.

Significant amounts of energy are flushed out of buildings in the water, air and compost streams. Off the shelf, on-site energy recycling technologies can effectively recapture energy from waste hot water and stale air and transfer that energy into incoming fresh cold water or fresh air. Recapture of energy for uses other than gardening from compost leaving buildings requires centralized anaerobic digesters.

HVAC systems are powered by motors. Copper, versus other metal conductors, helps to improve the electrical energy efficiencies of motors, thereby enhancing the sustainability of electrical building components.

Site and building orientation have some major effects on a building's HVAC efficiency.

Passive solar building design allows buildings to harness the energy of the sun efficiently without the use of any active solar mechanisms such as photovoltaic cells or solar hot water panels. Typically passive solar building designs incorporate materials with high thermal mass that retain heat effectively and strong insulation that works to prevent heat escape. Low energy designs also requires the use of solar shading, by means of awnings, blinds or shutters, to relieve the solar heat gain in summer and to reduce the need for artificial cooling. In addition, low energy buildings typically have a very low surface area to volume ratio to minimize heat loss. This means that sprawling multi-winged building designs (often thought to look more "organic") are often avoided in favor of more centralized structures. Traditional cold climate buildings such as American colonial saltbox designs provide a good historical model for centralized heat efficiency in a small-scale building.

Windows are placed to maximize the input of heat-creating light while minimizing the loss of heat through glass, a poor insulator. In the northern hemisphere this usually involves installing a large number of south-facing windows to collect direct sun and severely restricting the number of north-facing windows. Certain window types, such as double or triple glazed insulated windows with gas filled spaces and low emissivity (low-E) coatings, provide much better insulation than single-pane glass windows. Preventing excess solar gain by means of solar shading devices in the summer months is important to reduce cooling needs. Deciduous trees are often planted in front of windows to block excessive sun in summer with their leaves but allow light through in winter when their leaves fall off. Louvers or light shelves are installed to allow the sunlight in during the winter (when the sun is lower in the sky) and keep it out in the summer (when the sun is high in the sky). Coniferous or evergreen plants are often planted to the north of buildings to shield against cold north winds.

In colder climates, heating systems are a primary focus for sustainable architecture because they are typically one of the largest single energy drains in buildings.

In warmer climates where cooling is a primary concern, passive solar designs can also be very effective. Masonry building materials with high thermal mass are very valuable for retaining the cool temperatures of night throughout the day. In addition builders often opt for sprawling single story structures in order to maximize surface area and heat loss. Buildings are often designed to capture and channel existing winds, particularly the especially cool winds coming from nearby bodies of water. Many of these valuable strategies are employed in some way by the traditional architecture of warm regions, such as south-western mission buildings.

In climates with four seasons, an integrated energy system will increase in efficiency: when the building is well insulated, when it is sited to work with the forces of nature, when heat is recaptured (to be used immediately or stored), when the heat plant relying on fossil fuels or electricity is greater than 100% efficient, and when renewable energy is used.

Renewable Energy Generation

BedZED (Beddington Zero Energy Development), the UK's largest and first carbon-neutral eco-community
: the distinctive roofscape with solar panels and passive ventilation chimneys

Solar Panels

Active solar devices such as photovoltaic solar panels help to provide sustainable electricity for any use. Electrical output of a solar panel is dependent on orientation, efficiency, latitude, and climate—solar gain varies even at the same latitude. Typical efficiencies for commercially available PV panels range from 4% to 28%. The low efficiency of certain photovoltaic panels can significantly affect the payback period of their installation. This low efficiency does not mean that solar panels are not a viable energy alternative. In Germany for example, Solar Panels are commonly installed in residential home construction.

Roofs are often angled toward the sun to allow photovoltaic panels to collect at maximum efficiency. In the northern hemisphere, a true-south facing orientation maximizes yield for solar panels. If true-south is not possible, solar panels can produce adequate energy if aligned within 30° of south. However, at higher latitudes, winter energy yield will be significantly reduced for non-south orientation.

To maximize efficiency in winter, the collector can be angled above horizontal Latitude +15°. To maximize efficiency in summer, the angle should be Latitude -15°. However, for an annual maximum production, the angle of the panel above horizontal should be equal to its latitude.

Wind Turbines

The use of undersized wind turbines in energy production in sustainable structures requires the consideration of many factors. In considering costs, small wind systems are generally more expensive than larger wind turbines relative to the amount of energy they produce. For small wind turbines, maintenance costs can be a deciding factor at sites with marginal wind-harnessing capabilities. At low-wind sites, maintenance can consume much of a small wind turbine's revenue. Wind turbines begin operating when winds reach 8 mph, achieve energy production capacity at speeds of 32-37 mph, and shut off to avoid damage at speeds exceeding 55 mph. The energy potential of a wind turbine is proportional to the square of the length of its blades and to the cube of the speed at which its blades spin. Though wind turbines are available that can supplement power for a single building, because of these factors, the efficiency of the wind turbine depends much upon the wind conditions at the building site. For these reasons, for wind turbines to be at all efficient, they must be installed at locations that are known to receive a constant amount of wind (with average wind speeds of more than 15 mph), rather than locations that receive wind sporadically. A small wind turbine can be installed on a roof. Installation issues then include the strength of the roof, vibration, and the turbulence caused by the roof ledge. Small-scale rooftop wind turbines have been known to be able to generate power from 10% to up to 25% of the electricity required of a regular domestic household dwelling. Turbines for residential scale use are usually between 7 feet (2 m) to 25 feet (8 m) in diameter and produce electricity at a rate of 900 watts to 10,000 watts at their tested wind speed. Building integrated wind turbine performance can be enhanced with the addition of an aerofoil wing on top of a roof mounted turbine.

Solar Water Heating

Solar water heaters, also called solar domestic hot water systems, can be a cost-effective way to generate hot water for a home. They can be used in any climate, and the fuel they use—sunshine—is free.

There are two types of solar water systems- active and passive. An active solar collector system can produce about 80 to 100 gallons of hot water per day. A passive system will have a lower capacity.

There are also two types of circulation, direct circulation systems and indirect circulation systems. Direct circulation systems loop the domestic water through the panels. They should not be used in climates with temperatures below freezing. Indirect circulation loops glycol or some other fluid through the solar panels and uses a heat exchanger to heat up the domestic water.

The two most common types of collector panels are Flat-Plate and Evacuated-tube. The two work similarly except that evacuated tubes do not convectively lose heat, which greatly improves their efficiency (5%-25% more efficient). With these higher efficiencies, Evacuated-tube solar collectors can also produce higher-temperature space heating, and even higher temperatures for absorption cooling systems.

Electric-resistance water heaters that are common in homes today have an electrical demand around 4500 kW·h/year. With the use of solar collectors, the energy use is cut in half. The up-front cost of installing solar collectors is high, but with the annual energy savings, payback periods are relatively short.

Heat Pumps

Air-source heat pumps (ASHP) can be thought of as reversible air conditioners. Like an air conditioner, an ASHP can take heat from a relatively cool space (e.g. a house at 70 °F) and dump it into a hot place (e.g. outside at 85 °F). However, unlike an air conditioner, the condenser and evaporator of an ASHP can switch roles and absorb heat from the cool outside air and dump it into a warm house.

Air-source heat pumps are inexpensive relative to other heat pump systems. However, the efficiency of air-source heat pumps decline when the outdoor temperature is very cold or very hot; therefore, they are only really applicable in temperate climates.

For areas not located in temperate climates, ground-source (or geothermal) heat pumps provide an efficient alternative. The difference between the two heat pumps is that the ground-source has one of its heat exchangers placed underground—usually in a horizontal or vertical arrangement. Ground-source takes advantage of the relatively constant, mild temperatures underground, which means their efficiencies can be much greater than that of an air-source heat pump. The in-ground heat exchanger generally needs a considerable amount of area. Designers have placed them in an open area next to the building or underneath a parking lot.

Energy Star ground-source heat pumps can be 40% to 60% more efficient than their air-source counterparts. They are also quieter and can also be applied to other functions like domestic hot water heating.

In terms of initial cost, the ground-source heat pump system costs about twice as much as a standard air-source heat pump to be installed. However, the up-front costs can be more than offset by the decrease in energy costs. The reduction in energy costs is especially apparent in areas with typically hot summers and cold winters.

Other types of heat pumps are water-source and air-earth. If the building is located near a body of water, the pond or lake could be used as a heat source or sink. Air-earth heat pumps circulate the building's air through underground ducts. With higher fan power requirements and inefficient heat transfer, Air-earth heat pumps are generally not practical for major construction.

Sustainable Building Materials

Some examples of sustainable building materials include recycled denim or blown-in fiber glass insulation, sustainably harvested wood, Trass, Linoleum, sheep wool, concrete (high and ultra high performance roman self-healing concrete), panels made from paper flakes, baked earth, rammed earth, clay, vermiculite, flax linnen, sisal, seegrass, expanded clay grains, coconut, wood fiber plates, calcium sand stone, locally obtained stone and rock, and bamboo, which is one of the strongest and fastest growing woody plants, and non-toxic low-VOC glues and paints. Vegetative cover or shield over building envelopes also helps in the same. Paper which is fabricated or manufactured out of forest wood is supposedly hundred percent recyclable .thus it regenerates and saves almost all the forest wood that it takes during its manufacturing process.

Recycled Materials

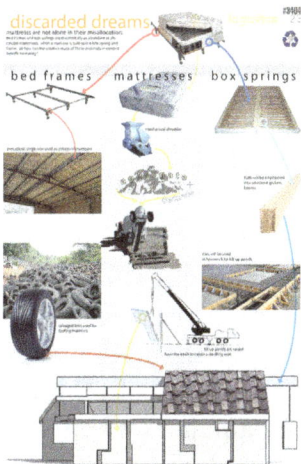

Recycling items for building

Sustainable architecture often incorporates the use of recycled or second hand materials, such as reclaimed lumber and recycled copper. The reduction in use of new materials creates a corresponding reduction in embodied energy (energy used in the production of materials). Often sustainable architects attempt to retrofit old structures to serve new needs in order to avoid unnecessary development. Architectural salvage and reclaimed materials are used when appropriate. When older buildings are demolished, frequently any good wood is reclaimed, renewed, and sold as flooring. Any good dimension stone is similarly reclaimed. Many other parts are reused as well, such as doors, windows, mantels, and hardware, thus reducing the consumption of new goods. When new materials are employed, green designers look for materials that are rapidly replenished, such as bamboo, which can be harvested for commercial use after only 6 years of growth, sorghum or wheat straw, both of which are waste material that can be pressed into panels, or cork oak, in which only the outer bark is removed for use, thus preserving the tree. When possible, building materials may be gleaned from the site itself; for example, if a new structure is being constructed in a wooded area, wood from the trees which were cut to make room for the building would be reused as part of the building itself.

Lower Volatile Organic Compounds

Low-impact building materials are used wherever feasible: for example, insulation may be made from low VOC (volatile organic compound)-emitting materials such as recycled denim or cellulose insulation, rather than the building insulation materials that may contain carcinogenic or toxic materials such as formaldehyde. To discourage insect damage, these alternate insulation materials may be treated with boric acid. Organic or milk-based paints may be used. However, a common fallacy is that "green" materials are always better for the health of occupants or the environment. Many harmful substances (including formaldehyde, arsenic, and asbestos) are naturally occurring and are not without their histories of use with the best of intentions. A study of emissions from materials by the State of California has shown that there are some green materials that have substantial emissions whereas some more "traditional" materials actually were lower emitters. Thus, the subject of emissions must be carefully investigated before concluding that natural materials are always the healthiest alternatives for occupants and for the Earth.

Volatile organic compounds (VOC) can be found in any indoor environment coming from a variety of different sources. VOCs have a high vapor pressure and low water solubility, and are suspected of causing sick building syndrome type symptoms. This is because many VOCs have been known to cause sensory irritation and central nervous system symptoms characteristic to sick building syndrome, indoor concentrations of VOCs are higher than in the outdoor atmosphere, and when there are many VOCs present, they can cause additive and multiplicative effects.

Green products are usually considered to contain fewer VOCs and be better for human and environmental health. A case study conducted by the Department of Civil, Architectural, and Environmental Engineering at the University of Miami that compared three green products and their non-green counterparts found that even though both the green products and the non-green counterparts both emitted levels of VOCs, the amount and intensity of the VOCs emitted from the green products were much safer and comfortable for human exposure.

Materials Sustainability Standards

Despite the importance of materials to overall building sustainability, quantifying and evaluating the sustainability of building materials has proven difficult. There is little coherence in the measurement and assessment of materials sustainability attributes, resulting in a landscape today that is littered with hundreds of competing, inconsistent and often imprecise eco-labels, standards and certifications. This discord has led both to confusion among consumers and commercial purchasers and to the incorporation of inconsistent sustainability criteria in larger building certification programs such as LEED. Various proposals have been made regarding rationalization of the standardization landscape for sustainable building materials.

Computer-aided Architectural Design

CAAD was introduced into the department of Architecture for more than a decade; however its full potentials have not been realized, therefore the benefits of employing CAAD tools in the design process is not enjoyed by students of architecture in the department. This situation can be linked to a plethora of problems such as: inadequate logistics for the teaching and practicing of CAAD, absences of CAAD training experts, complexities in the user interface of CAAD tools and on the frail creativity of the development work in CAAD. The current concept of architecture design education is a blend of both the traditional methods of drafting with the drawing board and T-square and the use of CAAD tools in the design process.

Design documents can be created by CAD tools in 2D, 2½D and 3D. 2D and 2½D designs can be used for standard building designs. 3D-software is needed for polyaxial curvatures, varying curves and dynamic structures. Drawing elements are vector-based such as points, lines, poly-lines, circles, etc. Textures simulate materials and depth. Complete building elements exist in specific libraries for the creation of models. These elements can be specified by various parameters. Linking single elements of this database forms a connected model with interdependent parts, which means that changes to one part of the model can influence other parts if necessary. Elevation drawings, sectional views and 3D views can be generated as construction drawings. The design data from these software tools can be exported and imported using various exchange formats for further use.

In an architectural design process, besides the physical tools such as all the drawing and drafting instruments, several conceptual tools such as the shape grammar and library are used. These conceptual tools are the ones that designers use to abstract and comprehend the design problem, mentally reconstruct, figure out and resolve it thus generate the design idea and the physical tools are the ones to visualize and realize the design.

From the above, physical tools like computers, contribute significantly to the realization of the final design scheme. Pioneers of CAAD designed the software to mimic the hardware tools such as: pencils, paper, and paint brush. Designing requires the designer to think visually and be creative, thus visual elements created during the design process influence the designers thinking. CAAD may act as a visual aid through 3D modeling and sketches. According to Asut , "Design is not a linear process which focuses on the target, but a netlike path which includes instantaneous feedbacks and coincidental decisions". Benton argues that the architectural designers must look beyond the complexities of the interface of CAAD tools as that leads to the development of predictive outcomes in design but rather view CAAD tools as "toys" that operate as speculative machines.

A designer should be solution- led but not problem led; use of physical tools of design helps in the evaluation of the solution through the BIM capability of physical tools such as Revit Architecture. The question at hand, is at what stage CAAD should be incorporated into the design process. One may argue that CAAD used in the design process becomes effective when the all the possible ideas of the task have been analyzed down to the smallest detail and the best solution arrived at; a view that results in the way CAAD is taught in the department, in this case the computer is reduced to a mere electronic board. However, recent advancement in CAAD tools makes it possible to incorporate CAAD-Systems at the conceptual stage of the design to the final stage of the design process.

CAAD programs are designed to have the capabilities of performing multiple complex design task with little human effort in a short time which otherwise would require intensive human effort and a lengthy period of time to accomplish with traditional methods of design, that is, manual drafting.

In the traditional method of drafting, any errors that needs to be corrected leads to a complete redrafting/retracing of the project. For instance, a spill of ink on the sheet or a change in the spatial arrangement of spaces on the floor plan requires redrafting of the project. In contrast designing with CAAD allows for the affected portion to be corrected without the need to redraw the entire project. CAAD programs are time-saver tools that allows for similar features in the design project to be repetitively drawn at their desired points. For instance, in a multi-storey building with typical floor plan, successive floor plans can be drawn by copying the initial plan and pasting at desired points. However, manual drafting requires redrawing each floor plan.

Active Design

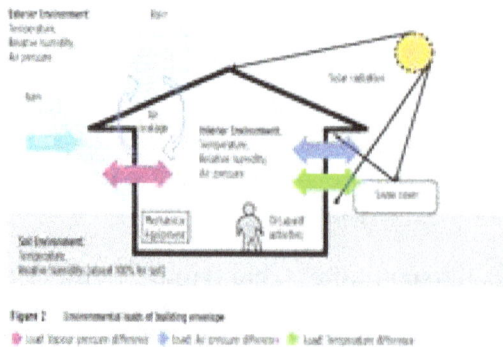

Figure 1   Environmental loads of building envelope

Building techniques use both active and passive design features in architecture to ensure comfortable living spaces, by means of utilizing energy intensive materials that enable overall reduction in energy usage.

Active designs use equipments that modify the state of the building, creating energy and comfort. While passive designs features are those that maximize energy efficiency by the actual design of the construction itself.

Sun being the main cause of heat gain through the effect, of solar energy transmission in buildings, it is necessary to understand its effects, which would enable designers to orient buildings properly and design shading devices.

Active architecture is the designs of buildings that contain mechanical devices, which transport the absorbed solar energy to other locations in the building. Active designs use equipments such as fans, air-conditioning, lights, pumps etc.

Selecting efficient equipment in active design, like using water conservation fixtures and appliances, choosing energy efficient appliances and lighting, providing, exhaust fans in bathrooms and the kitchen combined with a source of outside air are all means to an effective active design features in architecture.

The Ten Principles of Active Design:

1. Activity for all

   Neighbourhoods, facilities and open spaces should be accessible to all users and should support sport and physical activity across all ages.

   Enabling those who want to be active, whilst encouraging those who are inactive to become active.

2. Walkable communities

   Homes, schools, shops, community facilities, workplaces, open spaces and sports facilities should be within easy reach of each other.

   Creating the conditions for active travel between all locations.

3. Connected walking and cycling routes

   All destinations should be connected by a direct, legible and integrated network of walking and cycling routes. Routes must be safe, well lit, overlooked, welcoming, well-maintained, durable and clearly signposted. Active travel (walking and cycling) should be prioritised over other modes of transport.

   Prioritising active travel through safe, integrated walking and cycling routes.

4. Co-location of community facilities

   The co-location and concentration of retail, community and associated uses to support linked trips should be promoted. A mix of land uses and activities should be promoted that avoid the uniform zoning of large areas to single uses.

   Creating multiple reasons to visit a destination, minimizing the number and length of trips and increasing the awareness and convenience of opportunities to participate in sport and physical activity

5. Network of multifunctional open space

   A network of multifunctional open space should be created across all communities to support a range of activities including sport, recreation and play plus other landscape features including Sustainable Drainage Systems (SuDS), woodland, wildlife habitat and productive landscapes (allotments, orchards). Facilities for sport, recreation and play should be of an appropriate scale and positioned in prominent locations.

   Providing multifunctional spaces opens up opportunities for sport and physical activity and has numerous wider benefits.

6. High quality streets and spaces

   Flexible and durable high quality streets and public spaces should be promoted, employing high quality durable materials, street furniture and signage.

   Well designed streets and spaces support and sustain a broader variety of users and community activities.

7. Appropriate infrastructure

   Supporting infrastructure to enable sport and physical activity to take place should be provided across all contexts including workplaces, sports facilities and public space, to facilitate all forms of activity.

   Providing and facilitating access to facilities and other infrastructure to enable all members of society to take part in sport and physical activity

8. Active buildings

   The internal and external layout, design and use of buildings should promote opportunities for physical activity.

   Providing opportunities for activity inside and around buildings.

9. Management, maintenance, monitoring and evaluation

   The management, long-term maintenance and viability of sports facilities and public spaces should be considered in their design. Monitoring and evaluation should be used to assess the success of Active Design initiatives and to inform future directions to maximise activity outcomes from design interventions.

   A high standard of management, maintenance, monitoring and evaluation is essential to ensure the long-term desired functionality of all spaces.

10. Activity promotion and local champions

   Promoting the importance of participation in sport and physical activity as a means of improving health and wellbeing should be supported. Health promotion measures and local champions should be supported to inspire participation in sport and physical activity across neighbourhoods, workplaces and facilities.

   Physical measures need to be matched by community and stakeholder ambition, leadership and engagement.

Passive Design

'Passive design' is design that takes advantage of the climate to maintain a comfortable temperature range in the home. Passive design reduces or eliminates the need for auxiliary heating or cooling, which accounts for about 40% (or much more in some climates) of energy use in the average Australian home.

QMBA/Your New Home Magazine

The importance of passive design cannot be overstated. Paying attention to the principles of good passive design suitable for your climate effectively 'locks in' thermal comfort, low heating and cooling bills, and reduced greenhouse gas emissions for the life span of your home.

Passive design utilises natural sources of heating and cooling, such as the sun and cooling breezes. It is achieved by appropriately orientating your building on its site and carefully designing the building envelope (roof, walls, windows and floors of a home). Well-designed building envelopes minimise unwanted heat gain and loss.

The most economical time to achieve good passive design in a home is when initially designing and building it. However, substantial renovations to an existing home can also offer a cost effective opportunity to upgrade thermal comfort — even small upgrades can deliver significant improvements. If you're buying a new home or apartment, assess its prospects for thermal comfort and/or ability to be cost effectively upgraded to reflect good passive design principles in its climate.

For best results, 'passive' homes need 'active' users — people with a basic understanding of how the home works with the daily and seasonal climate, such as when to open or close windows, and how to operate adjustable shading.

A number of different and interrelated strategies contribute to good passive design. Passive design strategies vary with climate, as explained in more detail in Design for climate. The best mix of passive design strategies also varies depending on the particular attributes of your site. Choose a designer who is experienced in passive design for your climate and consider engaging a thermal performance expert to model different design options using thermal performance software.

Good passive design is critical to achieving a lifetime of thermal comfort, low energy bills and low greenhouse gas emissions.

Design for climate

Good passive design ensures that the occupants remain thermally comfortable with minimal auxiliary heating or cooling in the climate where they are built. Each of the eight main climate zones in Australia has its own climatic characteristics that determine the most appropriate design objectives and design responses. Identifying your own climate zone and gaining an understanding of the principles of thermal comfort helps you make informed design choices for your home. The Nationwide House Energy Rating Scheme (NatHERS), with its star classifications, is an additional and useful resource.

Orientation

Orientation refers to the way you place your home on its site to take advantage of climatic features such as sun and cooling breezes. For example, in all but tropical climates living areas would ideally face north, or as close to north as possible, allowing maximum exposure to the sun, and easy shading of walls and windows in summer. Good orientation reduces the need for auxiliary heating and cooling and improves solar access to panels for solar photovoltaics and hot water. Your home is thus more comfortable to live in and cheaper to run. It takes account of summer and winter variations in the sun's path as well as the direction and type of winds.

Shading

Shading of your house and outdoor spaces reduces summer temperatures, improves comfort and saves energy. Direct sun can generate the same heat as a single bar radiator over each square metre of a surface. Effective shading — which can include eaves, window awnings, shutters, pergolas and plantings — can block up to 90% of this heat. Shading of glass to reduce unwanted heat gain is critical, as unprotected glass is often the greatest source of heat gain in a house. However, poorly designed fixed shading can block winter sun. By calculating sun angles for your location, and considering climate and house orientation, you can use shading to maximise thermal comfort.

Passive solar heating

Passive solar heating is the least expensive way to heat your home. Put simply, design for passive solar heating keeps out summer sun and lets in winter sun while ensuring that the building envelope keeps that heat inside in winter and allows any built up heat to escape in summer. Orientation, thermal mass, sealing and other elements all contribute to the design of a house that benefits from passive solar heating. As most Australian climates require both passive heating and cooling, it's helpful along with Design for climate (to determine your climate zone) and Passive cooling.

Passive Cooling

Passive cooling is the least expensive way to cool your home. To be effective, passive cooling techniques need to cool both the house and the people in it — with elements such as air movement, evaporative cooling and thermal mass. Passive cooling design techniques can be applied to new homes as well as renovations, across a range of different climate zones. All Australian regions except those above the tropic of Capricorn require some form of passive heating in winter, so read this topic in conjunction with Design for climate and Passive heating.

Suntech Design

Sealing your Home

Air leakage accounts for 15–25% of winter heat loss in buildings and can contribute to significant loss of 'coolth' in climates where air conditioners are used. Sealing your home against air leaks is one of the simplest upgrades you can undertake to increase your comfort while reducing energy bills and greenhouse gas emissions. The more extreme your climate, the more beneficial sealing is, with the exception of naturally ventilated homes in the tropics. As sealing your home and increasing insulation levels can also create condensation and indoor air quality problems, this topic explains how condensation works, which climates present the greatest condensation risk and how you can limit its impact.

Insulation

Insulation acts as a barrier to heat flow and is essential for keeping your home warm in winter and cool in summer. It can also help with weatherproofing and soundproofing. A well-insulated and well-designed home provides year-round comfort, cutting cooling and heating bills by up to half and reducing greenhouse gas emissions. Climatic conditions determine the appropriate level of insulation as well as the most appropriate type to choose — bulk, reflective or composite. The most economical time to install insulation is during construction.

Insulation Installation

If insulation is to perform as intended then it must be correctly installed. For example, if bulk insulation is compressed, so are the air pockets within it that provide the insulation and it doesn't work effectively; neither does foil insulation if it is installed without an adjacent air gap. This topic explains, with illustrations, how to install insulation in a variety of construction types, and includes health and safety cautions, typical solutions and useful tips.

Thermal Mass

Thermal mass is the ability of a material to absorb and store heat energy. A lot of heat energy is needed to change the temperature of high density materials such as concrete, bricks and tiles: these materials have high heat storage capacity and are therefore said to have high thermal mass. Lightweight materials such as timber have low thermal mass.

Mike Cleaver, Clever Design

Use of materials with high thermal mass throughout your home can save significantly on heating and cooling bills, but thermal mass must be used appropriately. Poor use can exacerbate the worst extremes of the climate, radiating heat on a hot summer night or absorbing all the heat you produce on a winter night. Good use of thermal mass moderates indoor temperatures by averaging day–night temperature extremes. To be effective, thermal mass must be integrated with good passive design techniques appropriate for the climate. Although this is most easily done during construction or renovation, in many circumstances thermal mass can also be retrofitted.

Glazing

Glazed windows and doors bring in light and fresh air and offer views that connect interior living spaces with the outdoors. However, they can be a major source of unwanted heat gain in summer and heat loss in winter. Up to 40% of a home's heating energy can be lost and up to 87% of its heat gained through glazing. These thermal performance problems can be largely overcome by selecting the right glazing systems for your orientation and climate, and considering the size and location of window openings in your design. Use the Window Energy Rating Scheme (WERS), which rates the energy and energy-related performance of different window products.

Up to 40% of a home's heating energy can be lost and up to 87% of its heat gained through glazing.

Skylights

Skylights can make a major contribution to energy efficiency and comfort. They are an excellent source of natural light, perhaps admitting more than three times as much light as a vertical window of the same size, and can improve natural ventilation. However, they can be a major source of unwanted heat gain in summer and heat loss in winter. Factors to be considered when selecting from the many skylight options available include sizing and spacing (to control glare and heat gain), energy efficiency and appropriateness for climate.

Copper in Architecture

The usual grade of copper used for engineering, architectural and plumbing applications is phosphorus-deoxidized copper. It has a minimum copper content of 99.9 percent with a small addition of phosphorus, which allows this grade of copper to be welded and brazed. Its thermal conductivity, corrosion, heat and UV light resistance, ease of joining, high ductility, malleability, toughness and 100 percent recyclability (two thirds of the copper ever mined is still in use today) make copper the standard material for these applications. Electrical grade copper also has a greater than 99.9 percent copper content and is readily available in many forms.

Swanepoel explains: "Copper is one of very few metals that has a particular colour, meaning it is not simply silver-grey. Bright copper is reddish-pink, whilst oxidised copper is dark brown. Additionally when exposed to the elements, copper undergoes a change in colour – known as patination – transitioning from a reddish-pink to a blue-green."

Copper Alloys

"No metal is more suited to alloying than copper, a practice dating back to the beginning of civilisation, but still very much in use in many modern-day applications. Formed by mixing various compositions of metals in the molten state, alloys are used to expand properties for specific end uses. Alloys of copper give designers and architects further choices in terms of application and colour," continues Swanepoel.

Brasses are alloys of copper and zinc, and have a variety of attractive colours, ranging from red and yellow to gold and silver. With the addition of one percent manganese, brass will patinate to a chocolate-brown colour. Nickel silvers may be considered to be special brasses, and although they contain no silver content they resemble silver in appearance. Tin and phosphor bronzes are reddish-brown in colour; whilst aluminium bronzes have an attractive golden colour, which will darken slightly over time. Depending upon the copper content of copper-nickels, the colour of these alloys ranges from slightly pink-silver, to completely silver in colour, resembling the appearance of stainless steel.

With over 450 approved alloys to choose from, offering a wide range of properties and attributes, it is easy to select an appropriate alloy for the application and fabrication route required. In fact, there will usually be several that meet particular design and architectural requirements. Today, copper and copper alloys are available in sheets, meshes and expanded metal, which give an air of transparency.

Copper in the Home

Swanepoel explains: "The majority of us take for granted the lighting, heating, communications, running water, domestic appliances and entertainment systems in today's homes. All of these are reliant on copper components. In addition, copper and brass are widely used in both utility and decorative items such as cookware, door fittings and furniture. Copper also supports renewable energy and plays an essential role in the solar thermal heating and cooling systems, wind turbines and photovoltaic panels that are increasingly incorporated into 21st century homes."

Copper plumbing systems ensure the long lasting trouble-free and safe delivery of water for drinking, washing and heating. They are used in buildings of all types, from hotels and offices, to private houses and apartments. These buildings also remain secure thanks to locks and keys made from copper alloy components that ensure reliability, strength and freedom from corrosion.

Its excellent durability means that a copper component can often outlive the product or application of which it is a part. In buildings, copper cladding, a copper roof or copper guttering and downpipes can last for hundreds of years. Over time, copper used in outdoor applications will weather and oxidise and take on its familiar green patina. Manufacturers also now have factory methods that can apply oxidised or patinated surfaces straight away.

Science has demonstrated the naturally antimicrobial properties of copper in the fight against potentially life threatening infections. Bacteria and viruses, including those from the influenza family such as H_5N_1 (bird flu) and H_1N_1 (swine flu), are rapidly inactivated on contact with copper. This inherent property has seen the use of antimicrobial copper surfaces dramatically increase in hospitals and food preparation areas. Copper piping also helps limit the spread of Legionnaires' disease, as well as combating gastro-intestinal infections by reducing the risk of water being contaminated by the *Escherichia coli* or *Listeria* bacteria.

The Future—Breathable Buildings

With its pores and sweat glands, human skin might be one of world's best natural air conditioners. Biologist turned architect, Doris Kim Sung, proposed in her TED talk last October that building skins should be more similar to human skin. Considering that 30 to 40 percent of all primary energy consumed worldwide goes toward heating and cooling buildings, Sung's sustainable design concept could be a more passive method for ventilating buildings. The material she has in mind is a thermo-bimetal strip — two thin pieces of copper and steel sandwiched together. These two metals expand and contract when heated and cooled. When it is hot the metal bends one way, conversely when it is cold, it bends the other. This means that when direct sunlight hits the bimetal strips, they would bend inwards and close together to shade the building, then when the building gets too hot, the metal could bend in such way that opens up "pores" to release heat.

Climate-adaptive Building Shell

Improvements in design and construction of building shells plays an important role in recent efforts that aim to bridge the gap from current practice towards meeting our future energy saving targets. Notwithstanding the fact that good progress has been made, these attempts usually do not get around the status quo that building shells are typically designed as static elements in a dynamic environmental context. By being static or fixed, the conventional building shell has no means of responding to:

- The changes in weather conditions throughout the day and throughout the year.

- The variable nature of occupants' preferences.

In contrast, climate adaptive building shells (CABS) do offer the ability of actively moderating the exchange of energy across a building's enclosure over time. By doing this in a sensible way, in response to prevailing meteorological conditions and comfort needs, it introduces good energy saving opportunities. A growing interest in CABS therefore speculates on an added value on top of passive design solutions, and considers the concept as one of possible ways to accomplish the shift towards net zero energy buildings. The concept of CABS is referred to by a multitude of ambiguous terms, including: active, intelligent, dynamic, interactive, smart etc.

Current progress in the field of CABS is characterized by fragmented developments; either driven by specific advances in material science (e.g. switchable glazing, adaptable thermal mass and variable insulation), or originating from creative processes in design teams. Literature on CABS in relation to building performance simulation (BPS) shows the same degree of fragmentation, as it mainly deals with performance evaluations of specific case-studies such as: dynamic thermal insulation , and smart windows . Despite these efforts, it remains unclear what type of building envelope behaviour, actually results in the best building performance.

Within research settings, it has been demonstrated recurrently that the application of optimization techniques as a design aid, can move building performance beyond the level of "trial-and-error" designs. Initially, these developments led to the specification of generic design rules, derived on the basis of simplified building models. The advent of more efficient optimization algorithms, and the continuing trend of increasing computational power, now also enables optimization studies to be performed at a higher level of detail. In recent years, optimization was successfully deployed for design of building envelopes, adapted to specific conditions and contexts. For this purpose, researchers either work with tailor-made software, or use general-purpose optimization programs coupled to detailed building performance simulation tools, like EnergyPlus, ESP-r and TRNSYS. The next challenging step is to bring the power of optimization to practitioners, through development of user-friendly interfaces while respecting the aesthetics of architecture.

Since the role of optimization in CABS is thus far underexplored, the true value of making building shells adaptive is yet an unknown, and we can only guess how much of this potential is accessible with existing concepts and technologies.

Hurricane Resistant Building Design

Hurricane resistant building design protects a structure and its occupants from high winds, tornadoes, rain, and flooding. In hurricane prone regions, hurricane resistant design is essential; a category one hurricane can destroy mobile homes and damage roof, shingles, gutters, etc., but a category five hurricane (like Irma, with 157 mph and more winds) can destroy framed and mobile homes and cause total roof failure and wall collapse. The dangerous winds of a hurricane can also transform debris into flying missiles that can penetrate walls and threaten lives. However, flooding during a hurricane, which occurs due to storm surges, rain, and river overflow, is by far the biggest threat to life and property. For instance, in Louisiana during Hurricane Katrina, 40 percent of the 1577 deaths were from drowning. Best practices for hurricane resistant design must protect from surging water levels, pounding rains, and damaging winds for the duration of storms.

Best Practices for Flood Resistant Building Design

Best practices for hurricane resistant building design and construction in flood hazard zone must protect against flooding associated with storm surge and tide. Hurricane resistant building design must also protect against excessive rain. The design of a structure built in a flood hazard zone must be according to the American Society of Civil Engineers 24 (ASCE 24). The ASCE 24 is the referenced standard in the International Building Code® (IBC) and tells designers, architects, and builders the minimum requirements and expected performance for the design and construction of buildings and structures in flood hazard areas. Buildings designed according to ASCE 24 aim to resist flood loads and flood damage and complement the National Flood Insurance Program (NFIP) minimum requirements. Hurricane and flood resistant design in flood hazards zone should include elevated structures, materials that can get wet, and design assemblies that easily dry when exposed to moisture. Flood and water resistant design in flood hazard zones is essential in protecting a structure and the occupants during a hurricane event.

Best Practice for Wind Resistance Design

Best practice for hurricane resistant building design and construction must protect against strong wind and flying debris. A continuous load path is essential to holding a building together when high winds of a hurricane try to tear it apart. The continuous load path ensures that when a load, including lateral (horizontal) and uplift loads, attacks a building, the load will move from the roof, wall and other components toward the foundation and into the ground. A strong continuous load path is crucial to holding the roof, walls, floors, and foundation together during a hurricane event.

Insulated Concrete Blocks Create Hurricane Resistant Buildings

Buildings constructed with insulated concrete blocks (ICB) maintain their integrity during intense winds of a hurricane of over 200 mph. Buildings constructed of insulated concrete blocks are much stronger than steel-framed buildings and wood under extreme wind events. In fact, a study published by the Portland Cement Association (PCA), compared the structural load resistance of conventionally framed walls to insulating concrete form (ICF) walls. The study established that concrete walls have greater structural capacity and stiffness to resist the in-plane shear forces of high wind than steel or wood framed walls. The strength of concrete walls lessons the lateral twists and damage to non-structural elements of a building such as the electrical and plumbing. Utilizing insulated concrete blocks for hurricane-resistant construction can maintain a building's integrity during a strong wind event.

Insulated concrete blocks (ICB) also resist damage debris flying over 100 mph. A study by Texas Tech University compared the impact resistance of wind driven debris between conventionally framed walls and ICF walls. The study concluded that ICF walls resist the impact of wind driven hazards while conventionally framed walls didn't stop the penetration of airborne debris. Insulated concrete walls are the best protection from windblown debris to a building and its occupants during a hurricane event.

The Bautex Wall System Stands up to a Hurricane's Strength

The Bautex Wall System has the strength to resist the heavy winds and flying debris against even the strongest hurricanes like Harvey and Irma. Both hurricanes had peak wind speeds at landfall of over 130 miles per hour. The Bautex Blocks meet the Federal Emergency Management Agency FEMA 320 and FEMA 361 guidelines in storm zones with wind speeds up to 250 miles per hour. The Bautex Block has the strength and mass to resist the impact to wind driven debris at speeds greater than 100 mph. In addition to severe weather resistance, Bautex Blocks have the thermal performance required by the IRC and IBC and are fire-rated, noise-reducing, and easy to install. Bautex Walls are a good choice when designing for hurricane-resistant construction.

In today's climate, where more frequent and severe weather events are occurring due to global warming, it is essential that construction in flood hazard areas practice hurricane resistant design. Best practice for a hurricane resistant building design protects a building and its occupants from high winds, flying debris, flooding, and rain.

Storm surge is a rise of water generated by a storm, above the predicted tides. Storm tide is a water level rise due to both a storm surge and the astronomical tide. During a hurricane, storm surges and storm tides can cause extreme flooding in coastal areas. In fact, in 2008 Hurricane Ike's storm surge and heavy rains caused widespread damage to southeastern Texas, western Louisiana, and Arkansas; killing twenty people, with 34 others still missing. And, many of the lives lost during Hurricane Katrina occurred directly, or indirectly, as a result of storm surge.

Global warming refers to the modern day rise in global temperature near the earth's surface. The increase in temperature is due to increasing concentrations of greenhouse gases (carbon dioxide (CO_2), methane (CH_4), nitrous oxide (N_2O), and fluorinated gases) in the atmosphere. The explanation for global warming is straightforward.

The sun's energy falls on the earth as ultraviolet, visible (light), and infrared (heat) electromagnetic energy. The earth absorbs some of the sun's energy as thermal energy. The earth reflects another part of the sun's energy (infrared heat) back into the atmosphere where it either passes through the atmosphere or is reflected back to the earth's surface. Nitrogen and oxygen, which are the dominant gases in the atmosphere, allow infrared heat to pass through the atmosphere, while the greenhouse gases absorb infrared heat and redirect it back to the earth. The more greenhouse gases there are, the more heat is redirected back to earth; hence the increase in global temperatures near the earth's surface.

References

- McGrath, Brian (2013). Urban Design Ecologies: AD Reader. John Wiley & Sons, Inc. pp. 220–237. ISBN 978-0-470-97405-6

- Copper-in-architecture-and-interior-design: leadingarchitecture.co.za, Retrieved 29 March 2018

- Active-and-passive-designs-in-architecture-431: glazette.com, Retrieved 19 April 2018

- James, J.P., Yang, X. Indoor and Built Environment, Emissions of Volatile Organic Compounds from Several Green and Non-Green Building Materials: A Comparison, January 2004.[4] Retrieved: 2008-04-30

- The-impact-of-computer-aided-architectural-design-tools-onarchitectural-design-education-the-case-of-knu st-2168-9717-1000145: omicsonline.org, Retrieved 15 May 2018

- Sustainable-architecture, encyclopedias-almanacs-transcripts-and-maps: encyclopedia.com, Retrieved 25 March 2018

- Mark Jarzombek, "The Disciplinary Dislocations of Architectural History," Journal of the Society of Architectural Historians 58/3 (September 1999). Other articles in that issue by Eve Blau, Stanford Anderson, Alina Payne, Daniel Bluestone, Jeon-Louis Cohen and others

- Interior-design, Retrieved 15 May 2018

- Theory-of-architecture: britannica.com, Retrieved 18 July 2018

3

Building Construction

Construction is the process of building any infrastructure. It starts with planning, designing, financing and continues till the completion of the project. This chapter closely examines the key concepts of building construction, such as building code, construction 3D printing, steel frame and earthquake-resistant structures.

Construction

Construction means the building of something. This can mean the building of anything from motorways, to an office block, to a brand new cinema. It is happening all around us, all the time.

The bigger the project, the longer it will take to build, so some construction could last a few weeks, and some could last a few years. There is lots of hard work involved in constructing a building and there are various different stages.

During the entire construction cycle of a building for example, the work will start with the foundation stage and move through to the framing, exterior, drywall and the finishing stage. Depending on what is being built, these stages may differ and there may even be more or fewer stages.

Construction is a team project and requires a lot of working together. A construction project may require a very large or small workforce, depending on the size of the building project. No matter what the size is, there are lots of different jobs to do, requiring many different skills.

On a construction site, each person is responsible for doing their own job using their own skills. On one construction site there could be over 50 men and women all doing different jobs such as a roofer, engineer, electrician, plumber and many more. Not all the members of the workforce will be based on site either; some may be driving on the road delivering supplies or back at the office planning the construction.

Because there can be a lot of people working on a construction site at one time, there is a lot going on all at once. Building also requires many different types of equipment, ranging from small tools to large equipment such as vehicles. If not used correctly, both of these types of equipment can be very dangerous to the person using it and also everyone else around them. This is why safety is always the primary focus on a construction site.

Processes

Design Team

In the industrialized world, construction usually involves the translation of designs into reality. A formal design team may be assembled to plan the physical proceedings, and to integrate those proceedings with the other parts. The design usually consists of drawings and specifications, usually prepared by a design team including Architect, civil engineers, mechanical engineers, electrical engineers, structural engineers, fire protection engineers, planning consultants, architectural consultants, and archaeological consultants. The design team is most commonly employed by (i.e. in contract with) the property owner. Under this system, once the design is completed by the design team, a number of construction companies or construction management companies may then be asked to make a bid for the work, either based directly on the design, or on the basis of drawings and a bill of quantities provided by a quantity surveyor. Following evaluation of bids, the owner typically awards a contract to the most cost efficient bidder.

Shasta Dam under construction

The best modern trend in design is toward integration of previously separated specialties, especially among large firms. In the past, architects, interior designers, engineers, developers, construction managers, and general contractors were more likely to be entirely separate companies, even in the larger firms. Presently, a firm that is nominally an "architecture" or "construction management" firm may have experts from all related fields as employees, or to have an associated company that provides each necessary skill. Thus, each such firm may offer itself as "one-stop shopping" for a construction project, from beginning to end. This is designated as a "design build" contract where the contractor is given a performance specification and must undertake the project from design to construction, while adhering to the performance specifications.

Several project structures can assist the owner in this integration, including design-build, partnering and construction management. In general, each of these project structures allows the owner to integrate the services of architects, interior designers, engineers and constructors throughout design and construction. In response, many companies are growing beyond traditional offerings of design or construction services alone and are placing more emphasis on establishing relationships with other necessary participants through the design-build process.

The increasing complexity of construction projects creates the need for design professionals trained in all phases of the project's life-cycle and develop an appreciation of the building as an advanced technological system requiring close integration of many sub-systems and their individual components, including sustainability. Building engineering is an emerging discipline that attempts to meet this new challenge.

Financial Advisors

Trump International Hotel and Tower (Chicago)

May 23, 2006

September 14, 2007 (3 months before completion)

Construction projects can suffer from preventable financial problems. Underbids happen when builders ask for too little money to complete the project. Cash flow problems exist when the present amount of funding cannot cover the current costs for labour and materials, and because they are a matter of having sufficient funds at a specific time, can arise even when the overall total is enough. Fraud is a problem in many fields, but is notoriously prevalent in the construction field. Financial planning for the project is intended to ensure that a solid plan with adequate safeguards and contingency plans are in place before the project is started and is required to ensure that the plan is properly executed over the life of the project.

Mortgage bankers, accountants, and cost engineers are likely participants in creating an overall plan for the financial management of the building construction project. The presence of the mortgage banker is highly likely, even in relatively small projects since the owner's equity in the property is the most obvious source of funding for a building project. Accountants act to study the expected monetary flow over the life of the project and to monitor the payouts throughout the process. Cost engineers and estimators apply expertise to relate the work and materials involved to a proper valuation. Cost overruns with government projects have occurred when the contractor identified change orders or project changes that increased costs, which are not subject to competition from other firms as they have already been eliminated from consideration after the initial bid.

Large projects can involve highly complex financial plans and often start with a conceptual estimate performed by a building estimator. As portions of a project are completed, they may be sold, supplanting one lender or owner for another, while the logistical requirements of having the right trades and materials available for each stage of the building construction project carries forward. In many English-speaking countries, but not the United States, projects typically use quantity surveyors.

Legal Aspects

Construction along Ontario Highway 401, widening the road from six to twelve travel lanes

A construction project must fit into the legal framework governing the property. These include governmental regulations on the use of property, and obligations that are created in the process of construction.

When applicable, the project must adhere to zoning and building code requirements. Constructing a project that fails to adhere to codes does not benefit the owner. Some legal requirements come from malum in se considerations, or the desire to prevent indisputably bad phenomena, e.g. explosions or bridge collapses. Other legal requirements come from malum prohibitum considerations, or factors that are a matter of custom or expectation, such as isolating businesses from a business district or residences from a residential district. An attorney may seek changes or exemptions in the law that governs the land where the building will be built, either by arguing that a rule is inapplicable (the bridge design will not cause a collapse), or that the custom is no longer needed (acceptance of live-work spaces has grown in the community).

A construction project is a complex net of contracts and other legal obligations, each of which all parties must carefully consider. A contract is the exchange of a set of obligations between two or more parties, but it is not so simple a matter as trying to get the other side to agree to as much as possible in exchange for as little as possible. The time element in construction means that a delay costs money, and in cases of bottlenecks, the delay can be extremely expensive. Thus, the contracts must be designed to ensure that each side is capable of performing the obligations set out. Contracts that set out clear expectations and clear paths to accomplishing those expectations are far more likely to result in the project flowing smoothly, whereas poorly drafted contracts lead to confusion and collapse.

Legal advisors in the beginning of a construction project seek to identify ambiguities and other potential sources of trouble in the contract structure, and to present options for preventing problems. Throughout the process of the project, they work to avoid and resolve conflicts that arise. In each case, the lawyer facilitates an exchange of obligations that matches the reality of the project.

Interaction of Expertise

Apartment complex under construction in Daegu, South Korea

Design, finance, and legal aspects overlap and interrelate. The design must be not only structurally sound and appropriate for the use and location, but must also be financially possible to build, and legal to use. The financial structure must accommodate the need for building the design provided, and must pay amounts that are legally owed. The legal structure must integrate the design into the surrounding legal framework, and enforce the financial consequences of the construction process.

Procurement

Procurement describes the merging of activities undertaken by the client to obtain a building. There are many different methods of construction procurement; however the three most common types of procurement are traditional (design-bid-build), design-build and management contracting.

There is also a growing number of new forms of procurement that involve relationship contracting where the emphasis is on a co-operative relationship among the principal, the contractor, and other stakeholders within a construction project. New forms include partnering such as Public-Private Partnering (PPPs) aka private finance initiatives (PFIs) and alliances such as "pure" or "project" alliances and "impure" or "strategic" alliances. The focus on co-operation is to ameliorate the many problems that arise from the often highly competitive and adversarial practices within the construction industry.

Traditional

This is the most common method of construction procurement and is well established and recognized. In this arrangement, the architect or engineer acts as the project coordinator. His or her role is to design the works, prepare the specifications and produce construction drawings, administer the contract, tender the works, and manage the works from inception to completion. There are direct contractual links between the architect's client and the main contractor. Any subcontractor has a direct contractual relationship with the main contractor. The procedure continues until the building is ready to occupy.

Design-build

Construction of the *Phase-1* (first two towers) of the Havelock City Project, Sri Lanka

This approach has become more common in recent years, and also involves the client contracting a single entity that both provides a design and builds it. In some cases, the design-build package can also include finding the site, arranging funding and applying for all necessary statutory consents.

The owner produces a list of requirements for a project, giving an overall view of the project's goals. Several D&B contractors present different ideas about how to accomplish these goals. The owner selects the ideas they like best and hires the appropriate contractor. Often, it is not just one contractor, but a consortium of several contractors working together. Once these have been hired, they begin building the first phase of the project. As they build phase 1, they design phase 2. This is in contrast to a design-bid-build contract, where the project is completely designed by the owner, then bid on, then completed.

Kent Hansen pointed out that state departments of transportation usually use design build contracts as a way of progressing projects when states lack the skills-resources. In such departments, design build contracts are usually employed for very large projects.

Management Procurement Systems

In this arrangement the client plays an active role in the procurement system by entering into separate contracts with the designer (architect or engineer), the construction manager, and individual trade contractors. The client takes on the contractual role, while the construction or project manager provides the active role of managing the separate trade contracts, and ensuring that they complete all work smoothly and effectively together.

Management procurement systems are often used to speed up the procurement processes, allow the client greater flexibility in design variation throughout the contract, give the ability to appoint individual work contractors, separate contractual responsibility on each individual throughout the contract, and to provide greater client control.

In recent time, construction software starts to get traction – as it digitizes construction industry. Among solutions, there are for example: Procore, GenieBelt, PlanGrid, Bouw7, etc.

Types of Construction

In planning for various types of construction, the methods of procuring professional services, awarding construction contracts, and financing the constructed facility can be quite different. For the purpose of discussion, the broad spectrum of constructed facilities may be classified into four major categories, each with its own characteristics.

Residential Construction

Residential construction is the business of building and selling individual and multi-family dewellings. The market fragments into single-unit, manufactured, duplex, quad-plex, and apartments and condominiums. Manufactured housing further divides into mobile homes and pre-built houses. The business varies primarily in the size and scale of the operations. In the simplist form, a builder buys a peice of land, develops the land by clearing and grading it, and constructing roads, sidewalks, drainage, waste removal, electrical and water supplies. Then the builder offers to build either custom homes or pre-designed homes, or pre-manufactured homes, depending on the market he is attempting to serve. In certain instance the builder may build one or more homes on speculation or "spec" meaning that he builds the home without having a ready buyer on the hope that once the house is built, a buyer will appear.

The Residential Construction Environment

Each builder has to run something like a factory where the flow of product is fairly steady. Translated into builder's terms this means that the builder needs a ready supply of developed land, a pool of ready and available skilled and semi-skilled laborers, reliable suppliers who provide materials at competitive prices, working capital to cover labor, supply, and living expenses while the homes are under construction, and an approach to marketing his products. The more successful builders are able to keep at least one construction crew busy on a continuous basis. This means that at any given time three to six homes will be under construction. That way the foundation, framing, plumbing, wiring, HVAC, drywalling, cabinet making, trim carpentry, brick laying, painting, and cleaning crews can simply move from one house to the next. Builders who are unable to keep a complete collection of crews busy are forced to spend time trying to coordinate and schedule notoriously unreliable independent crews. Inevitably delays ensue either causing costs to creep up as a result of overtime, or overall delay of project completion.

Financing can also be a serious stumbling block. The housing industry is extremely cyclical. As a result, interest rates fluxuate, but so do lender's willingness to provide the interim financing a builder needs to stay in business. The ability to be realistic and forecast trends in interest rates and housing demand is a crucial skill if the builder is to avoid being stuck with a growing inventory of completed homes and consequential pressures from the bank.

Some builders gain flexibilty by building custom homes where the margins are greater. In this market, a few days delay, or a few cost overruns do not usually result in the builder taking a loss. Others gain flexibily by using uncommon materials and components. Sometimes these can result in higher costs but produce savings over time for the homeowner as in the case of better than normal installation. Or using new, lighter construction materials for roadways, for example.

Other builders move into remodeling and/or medical accomdation construction in periods of slackning demand.

Overlaying all of this are building codes established by towns, cities, and counties. Typically building inspections are fully paid for by the builder in permit fees that can run 3 to 4 percent of the price of the home. Inspectors seem to have their own timetables on getting out to approve sites so success often depends on building good relationships with the inspectors or developing political influence.

Keys to Success

Successful builders are those who can keep a steady stream of residences under construction. This allows for more predictability in the quality and availability of the needed labor. It also means that discounts can be obtained from suppliers who learn that they can count on a certain volume of business. Similarly, banks and lending institutions enjoy working with builders who are predictable in making their payments. As a result they are more willing to extend credit when it is needed. And successful builders build delays into their plans and schedules. They don't know if sickness or the weather will cause the delays but for sure something almost always does. Those who do it well also manage the expectations of their custoemrs. setting unrealistic expectations for completion can result in some very unpleasant consequences for buyer and builder alike. Quality is another key to success that should not be overlooked. Those who do it right the first time don't have to take money out of their profits to make things right. Maybe, in the final analysis, the residential home builder is the great communicator: needing to keep everyone up-to-date on the project status including, workers, sub-contractors, customers, banks, and building inspectors.

Institutional and Commercial Building

Institutional and commercial building construction encompasses a great variety of project types and sizes, such as schools and universities, medical clinics and hospitals, recreational facilities and sports stadiums, retail chain stores and large shopping centers, warehouses and light manufacturing plants, and skyscrapers for offices and hotels. The owners of such buildings may or may not be familiar with construction industry practices, but they usually are able to select competent professional consultants and arrange the financing of the constructed facilities themselves. Specialty architects and engineers are often engaged for designing a specific type of building, while the builders or general contractors undertaking such projects may also be specialized in only that type of building.

Because of the higher costs and greater sophistication of institutional and commercial buildings in comparison with residential housing, this market segment is shared by fewer competitors. Since the construction of some of these buildings is a long process which once started will take some time to proceed until completion, the demand is less sensitive to general economic conditions than that for speculative housing. Consequently, the owners may confront an *oligopoly* of general contractors who compete in the same market. In an oligopoly situation, only a limited number of competitors exist, and a firm's price for services may be based in part on its competitive strategies in the local market.

Specialized Industrial Construction

Specialized industrial construction usually involves very large scale projects with a high degree of technological complexity, such as oil refineries, steel mills, chemical processing plants and coal-fired or nuclear power plants. The owners usually are deeply involved in the development of a project, and prefer to work with designers-builders such that the total time for the completion of the project can be shortened. They also want to pick a team of designers and builders with whom the owner has developed good working relations over the years.

Although the initiation of such projects is also affected by the state of the economy, long range demand forecasting is the most important factor since such projects are capital intensive and require considerable amount of planning and construction time. Governmental regulation such as the rulings of the Environmental Protection Agency and the Nuclear Regulatory Commission in the United States can also profoundly influence decisions on these projects.

Infrastructure and Heavy Construction

Infrastructure and heavy construction includes projects such as highways, mass transit systems, tunnels, bridges, pipelines, drainage systems and sewage treatment plants. Most of these projects are publicly owned and therefore financed either through bonds or taxes. This category of construction is characterized by a high degree of mechanization, which has gradually replaced some labor intensive operations.

The engineers and builders engaged in infrastructure construction are usually highly specialized since each segment of the market requires different types of skills. However, demands for different segments of infrastructure and heavy construction may shift with saturation in some segments. For example, as the available highway construction projects are declining, some heavy construction contractors quickly move their work force and equipment into the field of mining.

Health and Safety Risks in Construction

The construction industry accident fatality rate stands at more than double that of the all sector average – more minor accidents are almost incalculably more. Put simply, construction sites are a health and safety nightmare – almost every conceivable hazard exists within this constantly changing working environment.

But the hazards associated with construction sites are well known – most responsible employers are aware of their duty of care to employees, visitors, and those that may be affected by their activities, and will manage the site effectively, implementing appropriate accident prevention measures. Risk assessments are carried out by management to identify hazards and risks posed.

Working at Height

The construction of buildings – or indeed, demolition works – frequently requires tradesmen to work at height. Fatalities and injuries involving height relating factors account for many accidents each year.

The risks associated with working at a height are often increased by added access and mobility restrictions. Training, including safety awareness training is essential for employees required to work at height.

Moving Objects

A construction site is an ever changing environment; hazards are inherent to this industry and only increase as a construction project progress, as things rise and expand.

Construction sites can get quite hectic what with the sheer volume of constantly moving vehicles and trades people – overhead lifting equipment shifting heavy loads, supply vehicles, dumper trucks everywhere, manoeuvring around a usually uneven terrain.

Slips, Trips and Falls

When you consider the diverse range of activities going on at a construction site at any one time it seems hardly surprising slips, trips, and falls happen on an almost daily basis.

Construction sites are a mish mash of holes in the ground, buildings at various stages of completion, scaffolding, stored materials and equipment: you really do need eyes in the back of your head at times.

Noise

Noise is a major hazard within the construction industry. Repetitive, excessive noise causes long term hearing problems and can be a dangerous distraction, the cause of accidents.

Beware, using simple ear plugs does not necessarily offer total protection against hearing damage – employers are required to carry out and document a comprehensive noise risk assessment – and issue appropriate PPE.

Hand Arm Vibration Syndrome

Hand arm vibration syndrome, or 'blue finger' as it is commonly referred to, is a painful and debilitating industrial disease of the blood vessels, nerves and joints, triggered by the prolonged use of vibratory power tools and ground working equipment.

This industrial disease is frequently cited in compensation claim cases opened by ex-construction workers who worked for years with little or no protection, using inappropriate and poorly maintained equipment.

Material and Manual Handling

Materials and equipment is being constantly lifted and moved around on a construction site, whether manually or by the use of lifting equipment. Different trades will involve greater demands, but all may involve some degree of risk.

Where employee's duties involve manual handling, then adequate training must be carried out. Where lifting equipment is used, then adequate training must also be carried out but may involve some form of test, to confirm competency. Records of training must be maintained for verification.

Collapse

Not exactly a hazard, more a risk – an accident in waiting.

Every year excavations and trenches collapse, bury and seriously injure people working in them – precautions need to be planned before the work starts.

The risk of an unintended collapse is generally more associated with demolition works or when a partially completed building or scaffolding collapses, but still accounts for a percentage of fatalities each year.

Asbestos

Today there is a new generation of construction workers, including; joiners, electricians and plumbers for whom asbestos is seen as a historical problem, something from the past that's now long gone but that is a mistakenly.

There are an estimated 500,000 public buildings in the UK that contain harmful asbestos materials: often hidden away, forgotten, and by and large, harmless – in its undisturbed state. Workers need to know where it is and what to do if they come across suspicious materials that might contain asbestos.

Airborne Fibres and Materials – Respiratory Diseases

Construction sites are a throng of activity and kick up a lot of dust an often invisible, fine, toxic mixture of hazardous materials and fibres that can damage the lungs, leading to diseases such as chronic obstructive pulmonary, asthma and silicosis.

Simply issuing PPE is not enough employers have a duty to ensure protective equipment is actually used. Failure to do so could render an employee to disciplinary action and in hot water with the health and safety executive.

Electricity

On average, three construction industry workers are electrocuted each year during refurbishment work on commercial and domestic buildings. People working near overhead power lines and cables are also at risk. There are also a growing number of electrocutions involving workers who are not qualified electricians but who are carrying electrical work, such as plumbers and joiners and decorators.

Building Construction

Building construction, the techniques and industry involved in the assembly and erection of structures, primarily those used to provide shelter.

Building construction is an ancient human activity. It began with the purely functional need for a controlled environment to moderate the effects of climate. Constructed shelters were one means by which human beings were able to adapt themselves to a wide variety of climates and become a global species.

Human shelters were at first very simple and perhaps lasted only a few days or months. Over time, however, even temporary structures evolved into such highly refined forms as the igloo. Gradually more durable structures began to appear, particularly after the advent of agriculture, when people began to stay in one place for long periods. The first shelters were dwellings, but later other functions, such as food storage and ceremony, were housed in separate buildings. Some structures began to have symbolic as well as functional value, marking the beginning of the distinction between architecture and building.

The history of building is marked by a number of trends. One is the increasing durability of the materials used. Early building materials were perishable, such as leaves, branches, and animal hides. Later, more durable natural materials such as clay, stone, and timber and, finally, synthetic materials such as brick, concrete, metals, and plastics were used. Another is a quest for buildings of ever greater height and span; this was made possible by the development of stronger materials and by knowledge of how materials behave and how to exploit them to greater advantage. A third major trend involves the degree of control exercised over the interior environment of buildings: increasingly precise regulation of air temperature, light and sound levels, humidity, odours, air speed, and other factors that affect human comfort has been possible. Yet another trend is the change in energy available to the construction process, starting with human muscle power and developing toward the powerful machinery used today.

The present state of building construction is complex. There is a wide range of building products and systems which are aimed primarily at groups of building types or markets. The design process for buildings is highly organized and draws upon research establishments that study material properties and performance, code officials who adopt and enforce safety standards, and design professionals who determine user needs and design a building to meet those needs. The construction process is also highly organized; it includes the manufacturers of building products and systems, the craftsmen who assemble them on the building site, the contractors who employ and coordinate the work of the craftsmen, and consultants who specialize in such aspects as construction management, quality control, and insurance.

Building construction today is a significant part of industrial culture, a manifestation of its diversity and complexity and a measure of its mastery of natural forces, which can produce a widely varied built environment to serve the diverse needs of society.

Building Construction Stages

Location

Locating land for constructing your dream home is very important. Identify the Land according to your choice. Always take the advice of an architect preferably with sound knowledge in Vastu and Kerala building rules to check and verify the plot which you have in mind. The Architect should be able to guide you considering Vastu Principles, soil conditions, and the latest building rules applying to your plot of land.

Verify with local authority office like Panchayath/municipality/Corporation whether there is any objection in building a house in that specific plot with respect to the survey no of your said plot. Also, make a point to check that your plot has no legal complications.

Design Process

Before finding/meeting up with an architect make a rough requirement list for your home. This process should cover the basic concepts of your home, such as living space, no: of bedrooms you want, Guest bedrooms, Hall, kitchen, Bathrooms, where you want the stairs, car porch, Garden/landscaping/Pond etc. You should involve your family also, ask their likes and suggestions. Visit your friends, talk to neighbours about pros and cons of their property and what they would like to improve.

Funding/Budgeting

How much will a house cost you to build? There has been a tremendous increase in the cost of house construction in Kerala. These days the normal running rate per square feet is anywhere between 1800 - 3500 + Rs including materials and labour. One should always make a note on the available cash at hand and also you may approach banks for loans in that way you come to know how much you can spare.

The Architect

Arrange a meeting with the architects. Talk through design goals and then fix an Architect, he will draw up schematics for consideration coupling functionality, room-by-room layouts, finding the best furniture positioning and use of space. Recommending structural changes where this seeks to make space work better. A 2D sketch is developed, which is then further refined and made to a 3D model, you can consult with the architect and his team to develop the final 3D Model of your house as well as an elevation. An approved architect will develop drawings for Panchayat/City municipality/Corporation to approve.

Building Permits

Permit to build your home is issued by the local governing bodies such as Panchayath/Municipality/Corporation which is normally validity for 2 years.

Building Contractor

Finding a good building Contractor is the next biggest task. And ideally, you have to identify a builder, plumber, mechanical, electrical contractor. But if you can find a good, reliable and reasonable contractor you can save your time and money. Always ask his method of working and check how reliable he is. You may visit and see some of his previous projects and speak with his previous clients.

You may visit and see some of his previous projects and speak with his previous clients. If he has good credentials, technical know-how and is also good at giving a quality work output then you can consider him as your Contractor.

Construction Process

Site Clearance - Before starting any construction work it becomes necessary to clear the place from the unwanted grass, boulder etc. In case of any hill like appearance on the ground, that too needs to be cleared of the excess earth and if there is a pit, it is required to be filled up. This total job is called site clearance.

Break Ground & Excavation - After the site clearance, the layout of the structure at the site can be planned with respect to the given foundation plans. Begin earth excavation and take trenches accordingly.

Foundation - A foundation is the lower portion of building a structure that transfers its gravity loads to the earth. Foundation work is done according to drawings provided by the Architect. i.e. the size of foundation, depth, length and breadth etc. and type of foundation (Rubble Packing or Raft and beams etc.)

Superstructure - Super-structure is to provide support in the construction of the building as per designed plan and various members of super-structure such as columns and beams are designed to provide strength for carrying the dead load and live load expected to come on the various parts of the structure in a safe and well-distributed manner. After casting the roofing slab necessary waterproofing coatings shall be done.

Stairs - Vasthu instructs and recommends to have rising stair steps going up in North-South direction or West-East direction if it is spiral stair steps going up in clockwise direction. But due to constraints on building design, it is not always possible to follow Vastu recommendations.

Boundary Wall and Gates- Compound walls should be built ideally just before beginning major construction activities, it is to protect the site and the material stored in the storage shed, from the outside environments and from thieves.

Roof/Heat Protective Coatings - truss work roofing, Weather Resistant Barrier, Waterproofing coatings, Rainscreen, Green roof are some of the applications you can implement to reduce heat.

Electrical And Lighting - works can be done after the masonry work has been completed.

Interior Design - works can be executed according to the working drawing provided by the architect.

Plastering

Plastering work can commence after the initial lighting and electrical plumbing work has been completed. Cement plaster is generally used with 13 mm thickness and sometimes it can be of 20 mm thickness. On completion of brickwork, plastering is to be done:

(a) to make the building structurally strong

(b) to protect it from the effect of weather, and

(c) to give it an attractive look.

Initial Plumbing - Once plastering is done for bathroom walls, it is okay to start plumbing works.

Painting - Is done with cement primer once initial wiring work and plumbing works are done.

Flooring

Flooring works can commence after the initial wiring works and primer coat is done to the interior walls. There are many types of floors according to their uses, economy and required the level of finishing. Ceramic tiles, Vitrified tiles, Clay Tiles, Granite, Marble, Wood, Epoxy flooring are some of the options you have in flooring.

Cabinets, Interior Works and Crockery Shelves

Interior works can commence after the initial wiring works and primer coat is done to the interior walls. Cabinets, shelves and kitchen can be done using a variety of materials which are currently available in the market such as Wood, Multi-Wood, MDF, Plywood, ACP, Stainless steel etc.

Finishing Plumbing Works - Can be done after tiling works are completed.

Finishing Electrical and Lighting works are done just before the application of final finishing coat of paint.

Completion Certificate

At this stage, one can apply for completion certificate from the respective authorities. After completion of construction Architect/Licensee will have to apply for a completion certificate in the prescribed format along with completed building drawings to the Issuing Authorities. The local authority will check completion documents for compliance with building rules and will assess building tax for the building. Once you have received the completion /occupancy certificate you can apply for water connection.

A temporary electrical connection can be applied any time once you have an active site which can be converted to a permanent connection later on when all the necessary papers are at hand.

Hardscaping and Landscaping

Once your builder has completed your home, there's still the "hardscaping" to be done – the driveway, patio and walkways, and then the "landscaping" plan can be put into action – the irrigation system, laying of grass bed and planting of trees, as well as outdoor lighting, to be considered.

Final Clean up

There will always be debris left over from the construction process on the interior and exterior of the home that you'll want to have removed/cleaned.

Building Code

Building codes specify minimum standards for the construction of buildings. The codes themselves are not legally binding. They serve, rather, as "models" for legal jurisdictions to utilize when developing statutes and regulations.

The main purpose of building codes are to protect public health, safety and general welfare as they relate to the construction and occupancy of buildings and structures.

The building code becomes law of a particular jurisdiction when formally enacted by the appropriate governmental or private authority.

Today, home and business construction has become process governed by a complex series of rules. A building code is usually not one document, but rather it is usually a series of documents setting forth requirements for several aspects of construction, such as gas, mechanics, electricity, fire-alarm systems, and plumbing. Building codes generally regulate all aspects of a construction project, including the structural design of a building, sanitation facilities, environmental control, fire prevention, ventilation, light, materials used for the building, and conservation measures. State and local governmental entities are empowered to enact building codes as part of their police powers under the Tenth Amendment to the federal Constitution. That amendment has been interpreted to allow the states to enact legislation designed to protect public health, welfare, and safety.

The development of modern building codes began in the early twentieth century. Residents who lived in tenement houses during that time began a movement that demanded basic sanitation in their housing. Insurance companies also advocated the use of safety standards, due to the potential limitations on the liability of these companies. In 1905, the National Board of Fire Examiners, the predecessor to the American Insurance Association, approved the first National Building Code.

It was designed to be used as a model by state and local governmental units when drafting their own building codes. This model code proved very popular among legislators because it provided a respected and comprehensive source for technical construction requirements without the burden and expense of researching and drafting a building code from scratch.

During the New Deal era of the 1930s, the federal government sought to modernize the system of housing in the United States, and the use of building codes to ensure safety and sanitation became widespread. Studies during the late 1960s and early 1970s indicated that the vast majority of cities had adopted a building code of some form. As the use of building codes became more prevalent, the actual codes themselves became much more comprehensive and complex. Through the 1970s, the majority of building codes were enacted at the local level.

A number of model building codes were developed during the second half of the twentieth century. By the 1990s, four major building codes were produced, including the National Building Code, by the American Insurance Association (AIA); the Basic/National Building Code (sometimes called the BOCA Code), by the Building Officials Conference of America (BOCA); the Southern Standards Building Code, by the Southern Building Code Congress International, Inc. (SBCCI); and the Uniform Building Code, by the International Congress of Building Officials (ICBO). Most of these various organizations were formed during the first half of the twentieth century by code enforcement officials who wanted to provide a forum whereby they could exchange ideas about the implementation of building codes.

During the past 20 years, roughly half of the states have enacted legislation providing construction standards on a statewide basis. The states that enacted these laws generally have done so in order to provide uniformity in building regulations across the state, and also to ensure that building laws protected all of the citizens in the state equally. Local governments have retained much of the responsibility for the actual implementation of building regulations in these states. It is not uncommon for a state to draft statutes that govern buildings on a general level, while the local units of the state enact more specific regulations to apply to that locality. Local building codes often remain uniform because these local governments typically rely upon one of the available model building codes.

The various associations representing code enforcement officials have formed broader associations for the purposes of collaboration. In 1972, BOCA, SBCII, and ICBO formed the Council of American Building Officials (CABO), which has successfully drafted such model codes as the CABO One and Two Family Dwelling Code and the CABO Model Energy Code. In 1994, the three major model code organizations formed the International Code Council (ICC), which has produced several international model codes. As of 2003, the ICC had developed more than a dozen international model codes, including the International Building Code. The ICC estimates that 46 states, plus the District of Columbia, Puerto Rico, and some federal agencies, enforce or have adopted at least one of the international codes. Building codes are directly affected by ongoing research regarding the performance of products, materials, or construction methods. Industry experts develop standards, which are documents that contain industry consensus regarding the methods by which the products, materials, or methods should be designed or employed. When an organization drafts a model building code, it typically refers to these standards in the text of the code. Since the standards are national in scope, the reference of these standards ensures that a local building code requires constructors to meet minimum national standards concerning details like safety and performance.

Few question that houses and other buildings are now designed to be much safer and more sanitary than were buildings constructed a century or longer ago, primarily as a result of the implementation of the various building codes throughout the United States. However, some commentators have noted that the requirements of these codes have caused construction prices to rise steadily, which in turn causes the costs of housing and other building usage to rise as well. Moreover, some critics maintain that the process of developing building codes is often as much of a process of negotiation between trade groups who are protecting their own interests as it is a completely scientific process.

Those who are involved in the drafting and implementation of building codes counter that building codes are designed with the health and safety of the public in mind. Results of testing performed during the development of standards are often readily available for inspection, so if questions of reliability arise, they often can be answered through a review of these testing procedures. Moreover, supporters note that state and local governmental entities are not bound to adopt the model building codes, and if a governmental unit disagrees with a provision in a model code, it is free to replace that provision with a requirement of its own creation. Accordingly, if a member of the public disagrees with a particular requirement, he or she generally may raise this issue with the appropriate governing body that decides whether a code or code provision should be adopted.

The model codes in the United States are currently developed by two organizations: the International Code Council (ICC) and the National Fire Protection Association (NFPA). Both organizations are on an 18-month code change cycle. The groups provide opportunities to introduce, review and comment, support or oppose, and challenge actions.

The national model codes may be adopted by state and local jurisdictions with or without modifications or amendments, depending on their needs. States and municipalities typically reserve the right to amend the model codes to assure that the requirements for design and construction of buildings are appropriate for the climatic, geographical, geological, political, and economic conditions within their jurisdiction.

The building code enacted or adopted via a legislative and/or regulatory process at the state or local level becomes the minimum legal requirements to which buildings are designed and constructed.

Model Codes

International Codes Council

The International Codes Council (ICC) publishes codes commonly referred to as the "I-Codes." Most states and local jurisdictions have opted to base their building codes on the I-Codes. PCA actively participates in the development process of the:

International Building Code (IBC) provides minimum requirements to safeguard the public health, safety and general welfare for all buildings except one- and two-family dwellings (town homes) not more than three stories above grade. Safeguards are to be provided through structural strength, means of egress facilities, stability, sanitation, adequate light and ventilation, energy conservation, and safety to life and property from fire and other hazards attributed to the build environment. Safety to fire fighters and emergency responders during emergency operations is also part of the IBC.

International Energy Conservation Code (IECC) regulates the design of building envelopes for adequate thermal resistance and low air leakage and the design and selection of mechanical, electrical, service water-heating and illumination systems and equipment which enables effective use of energy in new building construction.

International Fire Code (IFC) establishes regulations affecting or related to structures, processes, premises and safeguards regarding the hazard of fire and explosion arising from the storage,handling, or use of structures, materials, or devices; conditions hazardous to life, property or public welfare in the occupancy of structures; fire hazards in the structure or on the premises from occupancy or operation; and matters related to the construction, extension, repair, alteration, or removal of fire suppression or alarm systems.

International Residential Code (IRC) provides minimum requirements to safeguard the public health, safety, and general welfare for residential construction limited to one- and two-family dwellings (town homes) not more than three stories above grade. The intent is "to provide these safeguards through affordability, structural strength, means of egress facilities, stability, sanitation, light and ventilation, energy conservation and safety to life and property from fire and other hazards attributed to the built environment."

International Urban Wildland Interface (IUWIC) addresses buildings within urban-wildland areas, defined as geographical areas where structures and other human development meets or intermingles with wildland or vegetative fuels. The intent is to mitigate the risk of life and structures from intrusion of fire from wildland fire exposures and fire exposure from adjacent strurctures and to mitigate structure fires form spreading to wildland fuels.

Performance Code for Building and Facilities (PCBF) provides appropriate health, safety, welfare, and social and economic value, while promoting innovative, flexible and responsive solutions that optimize the expenditure and consumption of resources. PCBF provides an acceptable level of health, safety and welfare, and to limit damage to property from events that are expected to impact buildings and structures. It provides for an environment free of unreasonable risk of death and injury form fires; structure that will withstand loads associated with normal use and the severity associated with the location in which the structure is constructed; means of egress and access for normal and emergency circumstances; limited spread of fire both within the building and to adjacent properties; ventilation and sanitation facilities to maintain the health of occupants; natural light, heating, cooking and other amenities necessary for the well being of the occupants; and efficient use of energy. It also "establishes requirements necessary to provide an acceptable level of life safety and property protection from the hazards of fire, explosion or dangerous conditions in all facilities, equipment and processes."

At one time nearly every jurisdiction with a building code developed the code themselves. In order to gain some needed uniformity and to reduce administrative and development costs regional model codes were developed. While some states and jurisdictions continued to write their own codes, there were three regional model building codes:

- *National Building Code (NBC)* published by the Building Officials Conference of America (BOCA) used primarily in the Northeast;

- *Standard Building Code (SBC)* published by the Southern Building Code Congress, *International* (SBCCI) used primarily in the Southeast;

- *Uniform Building Code (UBC)* published by the International Conference of Building Officials (ICBO) used primarily west of the Mississippi River.

Each of the three model codes had different provisions addressing regional climatic, geologic, and political, and societal needs. For example, seismic design and construction provisions tended to be more advanced in the *Uniform Building Code (UBC)* which was used by states in very high seismic design categories. More advanced provisions for protection from hurricane and other high wind events tended to first appear in the *Standard Building Code (SBC)*. Cold weather and frost protection provisions tended to be addressed first in the *National Business Code (NBC)*.

In addition to these regional model codes there was a model residential and a model energy code published by the Council of American Building Officials (CABO):

- *Model Energy Code (MEC)*

- *One- and Two-Family Dwelling Code*

In 1997, the regional codes writing organizations and the Council of American Building Officials (CABO) agreed to cease publishing their codes and to have one series of national model building codes published by the International Codes Council (ICC).

The *National Electrical Code, Life Safety Code*, and other codes were and continue to be published by the National Fire Protection Association. Other organizations developed other model codes, such as the *Uniform Plumbing Code* and *Uniform Mechanical Code* published by the International Association of Plumbing and Mechanical Officials (IAPMO).

Construction 3D Printing

3D printing in construction, also known as contour crafting or building printing, is what many believe the future of construction. The printing of buildings has a lot of potential advantages when compared to conventional construction methods. There are lower labor costs involved, construction can be done quicker and there is less to no waste produced. The concrete waste and failed prints that do occur can be reused as raw material for printing.

This technique can also be used for the construction of extraterrestrial structures on for instance the Moon or other planets where environmental conditions are less conductive to human labor-intensive building practices.

Scale Models: The Humble Beginnings (1986-2000)

In 1984, Charles "Chuck" Hull invented stereolithography (SLA), a method of 3D printing where designers create a 3D model that is then printed layer by layer into a solid, physical object. The SLA process involves pointing a UV laser at liquid photopolymer which makes it solid.

The 3D printers most commonly used in the consumer sector are SLA printers, from companies like MakerBot and others. It was clear that this technique could have applications in manufacturing and beyond, so Chuck promptly took out some patents and founded 3D Systems, a 3D printing company that is still alive and well today.

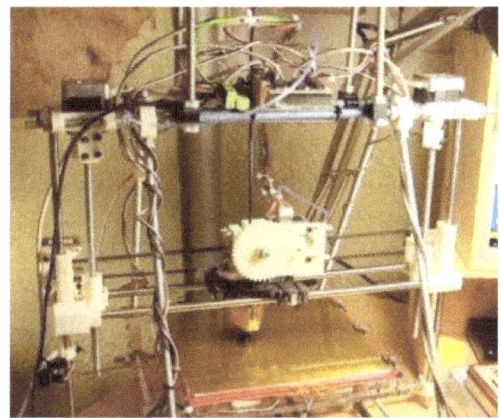

An early "homemade" 3D printer prototype.

One of the earliest uses for 3D printing was to print tabletop scale models for architecture firms. These models, in pre-BIM days, helped in the design process, and were valuable tools for both selling and planning building projects. 3D printing was a much cheaper way to erect these scaled-down models, compared to the time-intensive, hand-crafted replicas that had been the norm previously.

The advent of 3D and 4D Building Information Modeling (BIM) largely made the creation of physical models unnecessary, but many firms do still enjoy showing their work off this way.

By the 1990s, several organizations began experimenting with using 3D printing to produce modular components of full-scale projects. By the 2000s, these applications were in full swing and getting set to transform the entire industry.

Full-Scale Practical Applications (2000-2016)

In 2006, Dr. Behrokh Khoshnevis of the University of Southern California unveiled the Contour Crafting System, an enormous 3D printer designed to literally print buildings in place. It works like a desktop 3D printer, but uses a crane to do the printing, and concrete as the medium, to lay down a building's structural elements.

This Plastic Canal House: 2014

In 2014, a Dutch firm (DUS Architect) set out to demonstrate the potential for 3D printed architecture, by building a canal house out of 3D printed plastic in Amsterdam. The project uses a giant crane-like printing arm called the "Kamermaker," which literally means, "Room Builder." This project is ongoing.

This Steel Bridge: 2015

In 2015, the Dutch 3D Printing firm MX3D began printing a full-scale steel bridge, to be installed in downtown Amsterdam. When complete, the bridge will be fully functional. It is a proof of concept piece for the MX3D technology, which aims to make steel construction more cost efficient and faster.

Off-Grid Dwelling: 2016

In January 2016, architecture firm SOM announced a partnership with the US Department of Energy's Oak Ridge National Laboratory to produce highly efficient dwelling structures consisting of a 3D printed pod and a combination of renewable solar and natural gas energy systems. While the structures are not yet on the market, they promise to provide a cost-effective means of providing shelter in formerly inaccessible and remote locations, as well as sustainable long-term shelter for disaster relief.

The 3D printed shelter includes rooftop solar panels and an integrated battery system to power the dwelling day and night, and it comes with a companion vehicle that also generates its own power.

This Entire Concrete Mansion: 2016

In June 2016, Chinese company HuaShang Tenda announced that it had constructed an entire concrete mansion in 45 days. The company erected the building's frame first, placing plumbing and electrical wiring, and then printed the structure using 20 tons of inexpensive concrete and a computer-controlled printer. The two-story, 4,305 square foot dwelling claims to be earthquake-proof and environmentally friendly.

The Ten Houses in 24 Hours: 2016

In April 2016, another Chinese company claimed to have printed 10 houses in 24 hours. The walls are constructed from a mix of recycled construction material and cement, and the makers claim the homes are both cheaper and more environmentally friendly than traditional construction. The company also built its own office entirely out of 3D printed modules.

Design

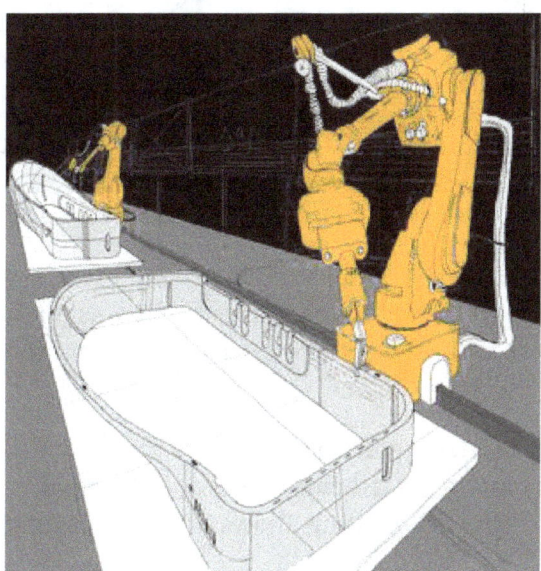

Free FAB Tower - Robotic production line of monocoque shells

Architect James Bruce Gardiner pioneered architectural design for Construction 3D Printing with two projects. The first Freefab Tower 2004 and the second Villa Roccia 2009-2010. FreeFAB Tower was based on the original concept to combine a hybrid form of construction 3D printing with modular construction. This was the first architectural design for a building focused on the use of Construction 3D Printing. Influences can be seen in various designs used by Winsun, including articles on the Winsun's original press release and office of the future The FreeFAB Tower project also depicts the first speculative use of multi-axis robotic arms in construction 3D printing, the use of such machines within construction has grown steadily in recent years with projects by MX3D and Branch Technology.

The Villa Roccia 2009-2010 took this pioneering work a step further with the a design for a Villa at Porto Rotondo, Sardinia, Italy in collaboration with D-Shape. The design for the Villa focused on the development of a site specific architectural language influenced by the rock formations on the site and along the coast of Sardinia, while also taking into account the use of a panellised pre-fabricated 3D printing process. The project went through prototyping and didn't proceed to full construction.

Francios Roche (R&Sie) developed the exhibition project and monograph 'We heard about' in 2005 which explored the use of a highly speculative self propelling snake like autonomous 3D printing apparatus and generative design system to create high rise residential towers. The project although impossible to put into practice with current or contemporary technology demonstrated a deep exploration of the future of design and construction. The exhibition showcased large scale CNC milling of foam and rendering to create the freeform building envelopes envisaged.

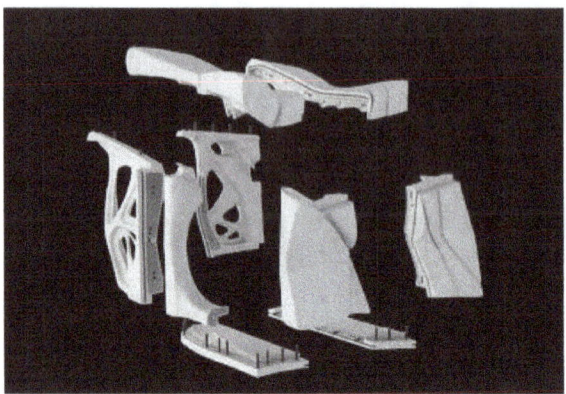

Villa Roccia - Detailed design exploded view

Dutch architect Janjaap Ruijssenaars's performative architecture 3D-printed building was planned to be built by a partnership of Dutch companies. The house was planned to be built in the end of 2014, but this deadline wasn't met. The companies have said that they are still committed to the project.

The Building On Demand, or BOD, a small office hotel 3D printed by 3D Printhuset's and designed by architect Ana Goidea, has incorporated curved walls and a rippling effects on their surface, to showcase the design freedom that 3D printing allows in the horizontal plane.

Structures

3D Printed Buildings

Europe's first residential 3D printed building

The BOD - 3D printed walls of the structure

The 3D Print Canal House was the first full-scale construction project of its kind to get off the ground. In just a short space of time, the Kamermaker has been further developed to increase its production speed by 300%. However, progress has not been swift enough to claim the title of 'World's First 3D Printed House'.

The first residential building in Europe and the CIS, constructed using the 3D printing construction technology, was the home in Yaroslavl (Russia) with the area of 298,5 sq. m. The walls of the building were printed by the company SPECAVIA in December 2015. 600 elements of the walls were printed in the shop and assembled at the construction site. After completing the roof structure and interior decoration, the company presented a fully finished 3D building in October 2017. The peculiarity of this project is that for the first time in the world the entire technological cycle of construction has been passed: design, obtaining a building permit, registration of the building, connection of all engineering systems. An important feature of 3D house in Yaroslavl, that also distinguishes this project from other implemented ones - this is not a presentation structure, but rather a full-fledged residential building. Today it is home of a real ordinary family.

Dutch and Chinese demonstration projects are slowly constructing 3D-printed buildings in China, Dubai and the Netherlands. Using the effort to educate the public to the possibilities of the new plant-based building technology and to spur greater innovation in 3D printing of residential buildings. A small concrete house was 3D-printed in 2017.

The Building on Demand (BOD), the first 3D printed house in Europe, is a project led by 3DPrinthuset for a small 3D printed office hotel in Copenhagen, Nordhavn area. As of November 2017, the building is in the final phase of applying fixtures and roofing, while all the 3DPrinted parts have been fully completed. The building is also the first 3D printed permanent building, with all permits in place and fully approved by the authorities.

3D Printed Bridges

3D printed bridge with the D-Shape technology. The first structure of this type in the world

In Spain, the first pedestrian bridge printed in 3D in the world (3DBRIDGE) was inaugurated 14th of December of 2016 in the urban park of Castilla-La Mancha in Alcobendas, Madrid. The 3DBUILD technology used was developed by ACCIONA, who was in charge of the structural design, material development and manufacturing of 3D printed elements. The bridge has a total length of 12 meters and a width of 1.75 meters and is printed in micro-reinforced concrete. Architectural design was done by Institute of Advanced Architecture of Catalonia (IAAC).

The 3D printer used to build the footbridge was manufactured by D-Shape. The 3D printed bridge reflects the complexities of nature's forms and was developed through parametric design and computational design, which allows to optimize the distribution of materials and allows to maximize the structural performance, being able to dispose the material only where it is needed, with total freedom of forms. The 3D printed footbridge of Alcobendas represented a milestone for the construction sector at international level, as large scale 3D printing technology has been applied in this project for the first time in the field of civil engineering in a public space.

Extraterrestrial Printed Structures

The printing of buildings has been proposed as a particularly useful technology for constructing off-Earth habitats, such as habitats on the Moon or Mars. As of 2013, the European Space Agency was working with London-based Foster + Partners to examine the potential of printing lunar bases using regular 3D printing technology. The architectural firm proposed a building-construction 3D-printer technology in January 2013 that would use lunar regolith raw materials to produce lunar building structures while using enclosed inflatable habitats for housing the human occupants inside the hardshell printed lunar structures. Overall, these habitats would require only ten percent of the structure mass to be transported from Earth, while using local lunar materials for the other 90 percent of the structure mass.

A rendering of the lunar base printing project, commissioned by the European
Space Agency in collaboration with Foster + Partners

The dome-shaped structures would be a weight-bearing catenary form, with structural support provided by a closed-cell structure, reminiscent of bird bones. In this conception, "printed" lunar soil will provide both "radiation and temperature insulation" for the Lunar occupants. The building technology mixes lunar material with magnesium oxide which will turn the "moonstuff into a pulp that can be sprayed to form the block" when a binding salt is applied that "converts this material into a stone-like solid." A type of sulfur concrete is also envisioned.

Tests of 3D printing of an architectural structure with simulated lunar material have been completed, using a large vacuum chamber in a terrestrial lab. The technique involves injecting the binding liquid under the surface of the regolith with a 3D printer nozzle, which in tests trapped 2 millimetres 0.079 in-scale droplets under the surface via capillary forces. The printer used was the D-Shape.

A variety of lunar infrastructure elements have been conceived for 3D structural printing, including landing pads, blast protection walls, roads, hangars and fuel storage. In early 2014, NASA funded a small study at the University of Southern California to further develop.

A 3D printed section of the lunar base made with the D-Shape technology,
as commissioned by the European Space Agency

The *Contour Crafting* 3D printing technique. Potential applications of this technology include constructing lunar structures of a material that could consist of up to 90-percent lunar material with only ten percent of the material requiring transport from Earth.

NASA is also looking at a different technique that would involve the sintering of lunar dust using low-power (1500 watt) microwave energy. The lunar material would be bound by heating to 1,200 to 1,500 °C (2,190 to 2,730 °F), somewhat below the melting point, in order to fuse the nanoparticle dust into a solid block that is ceramic-like, and would not require the transport of a binder material from Earth as required by the Foster+Partners, Contour Crafting, and D-shape approaches to extraterrestrial building printing. One specific proposed plan for building a lunar base using this technique would be called SinterHab, and would utilize the JPL six-legged ATHLETE robot to autonomously or telerobotically build lunar structures.

Advantages of 3D Printing in Construction

- Lower cost: 3D printing technology can save up to 60% of building materials and 50%-80% of manpower which improves work efficiency and helps to reduce the costs. This could be especially applicable and helpful in 3rd world countries, where better homes could be built for less cost.

- Environment friendly: 3D printing significantly reduces the amount of waste material created in production. The right amount of filament is used in creating the models each time, meaning you only use how much material you need. Many of the materials used to print objects can be made from recycled material and the created designs themselves can also be recycled.

- Fast delivery time: In an industry where construction delays can be extremely disruptive and costly, 3D printing offers new opportunities to accelerate delivery and reduce risk. By operating 24/7 and by reducing onsite glitches and hence delays, 3D printers can reduce construction times by 50%-70%.

- Design flexibility: With the help of 3D printers, architects are more flexible in the shape of their designs, no constrains of technical and structural aspects. They can realize even complex forms using non-linear shapes and curved walls which the conventional construction methods were not capable of building before.

Main Application Areas

3D printing has started to be used to print tabletop scale models for architecture companies before BIM (Building Information Modeling). It was an easy and fast method compared to traditional way which required more time and hand-craft. In the 1990s some companies started using 3D printing to produce modular components of full-scale projects. By the 2000s, these applications were in full swing and getting set to transform the entire industry. In the past years application areas of 3D printing has evolved a lot. You can 3D print building components, molds, entire buildings, bridges or interior design objects.

3D Printed villas in Shanghai by WinSun

So far the majority of applications were smaller-scale buildings and especially single-story houses except WinSun, a Chinese 3D printing company, which has developed the first continuous 3D printer for construction. The company printed the first batch of 10 houses in 2013. Using a special ink made of cement, sand and fiber, together with a proprietary additive, the printer adds layer by layer to print walls and other components in its factory. The walls are then assembled on site. Using up to 50% demolition waste and producing zero waste, the technology is environmental friendly. The impact on delivery time is even more impressive. Construction of a two-story 1,100 sqm villa take one day of printing, two days of assembly, with internal bar structures erected in advance, requiring three workmen only. According to the company 3D printing technology can save up to 60% of building materials and 50%-80% of manpower which improves work efficiency and helps to reduce the costs.

World's first 3D printed office building by WinSun

Winsun is also the first company to 3D print an office building which was opened in Dubai in May 2016. The entire structure was printed using a giant cement printer, then assembled on site. Printing took 17 days and was installed on in 2 days. Subsequent work on the building services, interiors, and landscape took approximately 3 months.

Buildings or building components are not the only areas of application. You can even print a bridge in metal. The most famous project is the one of the Dutch start-up MX3D who is 3D printing a fully functional stainless steel bridge to be installed on one of the oldest and most famous canals in the center of Amsterdam. MX3D equiped industrial multi-axis robots with 3D tools and developed the software to control them. What distinguishes this technology from traditional 3D printing methods is that the printing is done by 6-axis robot arms. This technique gives the design flexibility for architects and engineers and has huge potential to reduce the amount of material needed to make large structures. The printing of the pedestrian bridge is scheduled to be finalized early 2018.

Another application area of 3D printing in construction is the 3D printed molds which gives freedom to architects and designers to make their marks by unique shapes and forms. One of the well-known system in this area is the "FreeFAB system" a construction technique operated by the Australian-European contractor Laing O'Rourke. A giant robotic 3D printer, with a build volume of 30 x 3.5 x 1.5 meters, prints large molds from a specially designed wax; those molds are then used to cast concrete panels like the double-curved wall panels of London's Crossrail project. The technique is also more eco-friendly and less wasteful than conventional mold-making technologies. FreeFAB wax molds can be melted and the wax re-used again and again.

London Crossrail Project

3D printed wax molds

3D printing technology is evolving fast, we see impressive examples in the construction industry, however there is still a long way to go for most building constructions. First, the laws and authorizations must be modified according to this new technology. Also, 3D printers must be movable to be used anywhere. In short term we can expect more and more building components to be built with 3D printing as well as printed molds. It's sure that the buildings of future will not look like the ones of today. Architects, engineers and workers will have to adapt to this transformation and learn to master well these techniques of 3D printing. Even though there are still some challenges to overcome we can still be positive about the future of construction 3D printing.

Steel Frame

The important feature of steel framing is its flexibility. It can bend without cracking, which is another great advantage, as a steel building can flex when it is pushed to one side by say, wind, or an earthquake. The third characteristic of steel is its plasticity or ductility. This means that when subjected to great force, it will not suddenly crack like glass, but slowly bend out of shape. This property allows steel buildings to bend out of shape, or *deform*, thus giving warning to inhabitants to escape. Failure in steel frames is not sudden - a steel structure rarely collapses. Steel in most cases performs far better in earthquake than most other materials because of these properties.

However one important property of steel is that it quickly loses its strength in a fire. At 500 degrees celsius (930 degrees F), mild steel can lose almost half its strength. This is what happened at the collapse of the World Trade Towers in 2001. Therefore, steel in buildings must be protected from fire or high temperature; this is usually done by wrapping it with boards or spray-on material called *fire protection*.

Where Steel Frame Structures are used

Steel construction is most often used in

- High rise buildings because of its strength, low weight, and speed of construction
- Industrial buildings because of its ability to create large span spaces at low cost
- Warehouse buildings for the same reason
- Residential buildings in a technique called *light gauge steel construction*
- Temporary Structures as these are quick to set up and remove

Different types of structural steel framing systems are as follow:

- Skeleton steel framing
- Wall bearing steel framing
- Long span steel framing

Skeleton Steel Framing System

Skeleton steel frame is composed of steel beams and columns which are connected using proper connection. Steel beams around perimeter of the structure is termed as spandrel beams on which masonry walls are placed.

Typical plane views of skeleton framing are shown in below Figure and constructed skeleton steel frame are shown in figure. Steel columns, primary and secondary steel beams are shown in the figures.

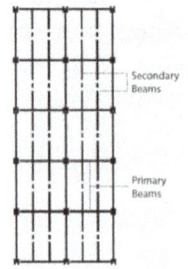
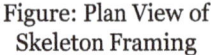

Figure: Plan View of   Figure : Skeleton Steel Framing
Skeleton Framing

There are different types and configurations of steel connections which are used to connect steel beams to columns in skeleton frame structure, for example, bolt connection and welded connections.

Figure-illustrate various types of bolted connection including flexible end plate, fin plate and double angle cleat.

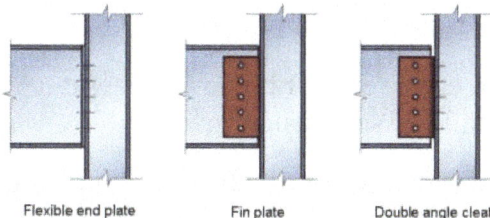

Flexible end plate   Fin plate   Double angle cleat

Figure : Different Bolt Connection Arrangement used to Connect Beams to Columns

It should be known that all gravity loads in skeleton frame structure are supported by beams and columns. The distance between columns can be established according to the functions and requirements of the structure.

Therefore, there are no restrictions that limit the area of the floor and roof of the building. Multi storey structures are possible to construct using skeleton framing.

Wall Bearing Steel Framing System

In a wall bearing steel framing structure, building wall whether it is interior or exterior is used to carry the end of structural members that support floor or roof loads.

Wall bearing should be adequately strong to not only be able to carry vertical reactions but also to resist any imposed horizontal loads.

Wall bearing framing is suitable for the construction of low rise structure. This is because the size of the bearing wall must be increased significantly to withstand considerably loads exerted in the case of multistory buildings.

This problem might be solved to certain extent if the reinforced concrete walls are applied.

There are several cases in which wall bearing frame system is suitable to be used. For example, single story house in which steel beams are used to carry wall and floor loads and the end of the steel beams are placed on foundation walls, as shown in figure below.

A further application of wall bearing system application is the utilization of steel beams known as lintels over wall openings like doors and windows.

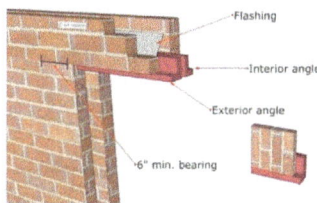

Figure: Using Steel Beam to Support Masonry Lintels

Figure : The end of steel beam, which support floor loads, is installed on walls, intermediary support, applied to support the beam because the span is large

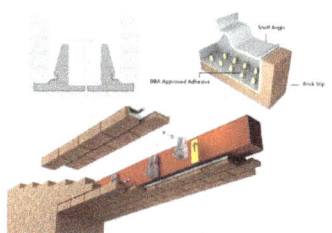

Figure: Using Steel Beam to Support Masonry Lintels

Long Span Steel Framing

Long span steel framing is considered when large clearance is required and such long spanning cannot be realized using steel beams and columns.

Long span steel framing options can be categorized into different types, for instance, girders, trusses, rigid frames, arches and cantilever suspension spans.

These classes of long span steel framing options along with their applications and various configurations are provided in table.

Table: Long Span Steel Framing Types, their Applications and Various Configurations

Long span framing system types	Application condition	Various types or configuration of the given long span framing system class
Girders	It is selected for the case where depth of the member is restricted over a large unobstructed area in lower storeys. The girder should support loads from above storeys across cleared area.	Roller beams, Riveted girder, welded girder, heavy girder, hybrid girder, and a girder consist of two girder fastened
Truss, Figure	It can be used for the case where restriction on the element depth is not imposed. it an economical way of spanning long distances provided that depth limitation is not existed. Trusses are better compare with other option in controlling deflections due to better stiffness.	Pratt, warren, fink, scissor, bow string and Virendeel
Arches, Figure	It is used to carry walls and roof with open or solid web arches	Hinge less arches, two hinged arches and three hinged arches
Rigid frames, Figure	It is used to span long distances. it is aesthetically pleasing that is why applied in the construction of churches, gymnasiums, auditoriums, bowling alleys and shopping centers	Single span rigid frame, welded rigid frame.

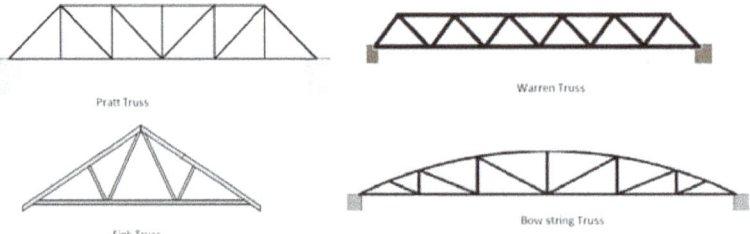

Figure : Different types of trusses used in steel structure construction

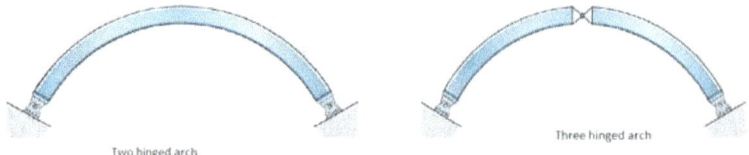

Figure : Types of Arches Used for Steel Structure Construction

Figure : Single Span Rigid Frame

Advantages of Steel Frame Structure

There are multiple reasons why steel makes an attractive building option from start to finish, not the least of which are:

- Sustainability

- Affordability

- Durability

From full-throttle metal building projects, to hybrid construction projects that leverage the attributes of both wood and steel, the modern builder has a wealth of options to choose from.

Steel is a versatile building material, which has led to its inclusion in nearly every stage of the construction process from framing and floor joists, to roofing materials. Here are some of the main benefits that make structural steel such a reliable choice.

Steel is Lighter than Wood

This may seem surprising at first, because if you weigh a 2x4 of wood and a 2x4 of steel, the steel will weigh more as the result of its density. When it comes to framing, however, The design of a steel I-beam will almost always cause it to be lighter than the lightest, structurally sound wood beam design. A steel I-beam weighs less than glulams, LVL, and Parallam beams.

In addition to decreasing the labor required to build with steel, the lighter-weight advantage reduces materials shipping costs, and can also simplify the design of a building's foundation and other structural support systems, which can further reduce project budgets.

You Can Build Faster with Steel

Time has always equaled money, but it seems like this high-tech era of ours has made it so that every clients wants their building to come in under budget and ahead of schedule. Fast-tracked projects can be a nightmare for architects and construction crews - namely because taking shortcuts can lead to unsafe building practices and a finished product that isn't up to snuff. That all changes with the addition of steel.

Steel parts are pre-engineered to a specific design inside the manufacturing plant and are shipped out, ready to be erected. This speeds up construction time significantly, making it possible to complete large-scale projects in a matter of weeks.

Because the fabrication process is highly quality-controlled, project managers can place their attention on other issues and the pre-cut, ready to assemble parts eliminate the need for measuring and cutting on site. This also takes the element of human error out of the equation, reducing the amount of time spent assembling something only to find out it needs to be re-measured, cut and installed again.

In addition to project time and budget issues, a faster construction timeline also reduces the amount of time your construction project impedes traffic, affects the flow into and out of surrounding businesses and any water or utility disruptions to nearby buildings.

Save Money with Steel

Much of the cost savings you'll gain can be inferred from the labor and cost benefits of decreased construction time. However, building with steel also saves money via other first time and lifetime savings.

- Steel can be recycled. Rather than paying landfill fees for non-recyclable construction waste, your company will be able to recycle steel and metal building components. Due to public interest in decreasing unnecessary constuction waste, most waste removal companies have subsidized programs allowing them to pick up your steel and metal building waste at no cost to you.

- Because steel is so durable, and requires so little maintenance, it is a more economic choice for building owners. Maintenance fees, repairs and replacements are minimal - even over the course of 50 years or more - saving building owners tens of thousands of dollars over the course of the building's lifetime.

- Innovation in steel production, combined with greater competition to meet rising steel demands, has brought steel prices lower than they've been in twenty years. According to the American Institute of Steel Construction, "In 1980, 10 man-hours were required to produce a single ton of steel. Today that same ton of structural steel requires substantially less than a single man-hour." Thus, these cost savings can be being passed on to the consumer.

- Due to a steel structures' almost unrivaled ability to withstand high winds, heavy snow loads, fire and seismic activities, combined with their resistance to pests and decay, insurance companies often offer lower premiums on policies underwritten for metal buildings.

- Faster construction times means fewer interest payments to the lender, who typically requires that interest-payments are made through the duration of the construction process.

When bundled together, these cost-saving benefits make steel one of the most affordable building products on the market.

Steel is Incredibly Versatile

The versatility of steel is unrivaled. From the ability for structural steel to be molded into virtually any shape to its exterior ability to yield shingle-esque roofing patterns and wood-like siding, steel's versatility is part of what is making it such an attractive option for the residential construction market.

Architects and designers like steel's ability to let their artistic imaginations run wild, while still having the ability to design and construct a building that is both safe and resilient. This same durability is also what allows for the versatile design of large, clear span buildings such as airplane hangers, warehouses, agricultural buildings and indoor arenas. It also permits for the construction of skyscrapers, the tallest of which stands in Dubai at 2722.4 feet (829.8 m) tall. The commercial sector no longer corners the market on steel buildings, either.

This same design versatility and flexibility is now being touted in the residential sector as well. Consider a family who wants to knock out a wall for a remodel or renovation, only to find that a load-bearing wood pillar is an essential component. Now, they have the option of running a steel beam across the ceiling, opening the space up and negating the need for a structural beam below the ceiling line. Additionally, steel and metal are used for siding and roofing materials that far outlast their wood counterparts.

Steel and Wood Hybridization

This is an area where hybridization comes into play as well. While most homeowners like the idea of steel's versatility and durability, they are wary of moving away from wood completely. As a result, many construction companies offer hybrid steel/wood buildings, which take advantage of the strength, durability and support benefits offered by steel, paired with the insulatory properties of wood.

This pairing of materials is so beneficial that the US Department of Housing and Urban Development has published a Hybrid Wood and Steel Details Building Guide to encourage urban developers to use structural steel in addition to traditional building materials.

It's Environmentally Friendly

Steel is made from recycled materials and can be recycled at the end of its lifespan, one of the many reasons why it can earn builders points toward major green building certification programs. According to the Steel Recycling Institute:

- 80 million tons of steel are recycled each year, making it the world's most recycled product.

- Since 1990, the steel industry has reduced energy intensity per ton of steel produced by 28% and CO_2 emissions by 35% per ton of steel shipped.

- Reductions in energy use and CO_2 emissions are rapidly reaching the limits defined by the laws of physics.

When combined with other design enhancements, steel buildings are incredibly energy efficient. The connections between high-quality, prefabricated steel parts is so exact that with the addition of adequate insulation, they are air-tight and comfortable, ensuring the building has a completely sealed envelope. Roof panels are primed and ready to host a solar array and cool metal roofing products dramatically decrease solar heat gain, further increasing energy savings.

Disadvantages of Steel Frame Structure

- Maintenance cost of a steel structure is very high. Due to action of rust in steel, expensive paints are required to renew time to time. So that resistance against severe conditions increases.

- Steel has very small resistance against fire as compared to concrete. Almost from 60°-70°C half of steel strength reduced.

- Steel cannot be mold in any direction you want. It can only be used in forms in which sections originally exists.

- If steel loses its ductility property, than chances of brittle fractures increase.

- If there are very large variations in tensile strength than this lead steel to more tension. Due to which steel tensile properties graph falls down.

Earthquake-resistant Structures

Earthquake-resistant construction, the fabrication of a building or structure that is able to withstand the sudden ground shaking that is characteristic of earthquakes, thereby minimizing structural damage and human deaths and injuries. Suitable construction methods are required to ensure that proper design objectives for earthquake-resistance are met. Construction methods can vary dramatically throughout the world, so one must be aware of local construction methods and resource availability before concluding whether a particular earthquake-resistant design will be practical and realistic for the region.

There is a fundamental distinction between the design of a building and the construction methods used to fabricate that building. Advanced designs intended to withstand earthquakes are effective only if proper construction methods are used in the site selection, foundation, structural members, and connection joints. Earthquake-resistant designs typically incorporate ductility (the ability of a building to bend, sway, and deform without collapsing) within the structure and its structural members. A ductile building is able to bend and flex when exposed to the horizontal or vertical shear forces of an earthquake. Concrete buildings, which are normally brittle (relatively easy to break), can be made ductile by adding steel reinforcement. In buildings constructed with steel-reinforced concrete, both the steel and the concrete must be precisely manufactured to achieve the desired ductile behaviour.

Building failures during earthquakes often are due to poor construction methods or inadequate materials. In less-developed countries, concrete often is not properly mixed, consolidated, or cured to achieve its intended compressive strength, so buildings are thus extremely susceptible to failure under seismic loading. This problem is often made worse by a lack of local building codes or an absence of inspection and quality control.

Building failures are also frequently attributed to a shortage of suitable and locally available materials. For instance, when a building is designed with steel-reinforced concrete, it is critical that the amount of steel used is not reduced to lower the building cost. Such practices substantially weaken a building's ability to withstand the dynamic forces of an earthquake.

Under normal conditions, a building's walls, columns, and beams primarily experience only vertical loads of compression. However, during an earthquake, lateral and shear loading occurs, which results in tensile and torsional forces on structural elements. Those forces result in high stresses at the building's corners and throughout various joints.

Strong construction joints are critical in building a structure that will withstand the shear loading of an earthquake. Since stress is concentrated at the joints between the walls, it is important that all the joints be properly prepared and reinforced. Concrete joints must also be properly compacted and anchored in order to achieve optimum strength. In the case of unreinforced masonry joints (mortar joints, such as those found in brick buildings), the anchoring between adjacent walls is especially important. When all the joints are tied together well, the building will act as a single integrated unit, enabling the forces of an earthquake to be transferred from one section to the next without catastrophic failure.

Earthquake-resistant construction requires that the building be properly grounded and connected through its foundation to the earth. Building on loose sands or clays is to be avoided, since those surfaces can cause excessive movement and nonuniform stresses to develop during an earthquake. Furthermore, if the foundation is too shallow, it will deteriorate, and the structure will be less able to withstand shaking. The foundation should therefore be constructed on firm soil to maintain a structure that settles uniformly under vertical loading.

To earthquake-proof buildings, engineers must ensure that the structures and their foundations are resistant to potential horizontal loads by employing a number of key design features:

Diaphragms

Diaphragms are the primary component of a building's horizontal structure, including the floors and the roof. To earthquake-proof a building, diaphragms must be placed on their own deck and strengthened horizontally to share forces with vertical structures.

Trusses

Trusses strengthen the diaphragm where the deck is weakest. Simply put, they are diagonal structures that are inserted into the rectangular areas of the frame.

Cross-bracing

Engineers incorporate a variety of columns, braces, and beams to transfer seismic forces back to the ground. Cross braces incorporate two diagonal sections in an X-shape to build wall trusses.

Shear Walls

To help resistance swaying forces, engineers use vertical walls, known as shear walls, to stiffen the structural frame of the building. These can be used in place of braced frames or in addition to them.

Moment-resisting Frames

Since shear walls limit a building's flexibility, some designers choose moment-resisting frames to allow positive movement. Although columns and beams can bend, joints and connectors stay rigid. These features also give building designers more flexibility to create exterior walls, ceilings, and arrange building contents.

A light Roof

As a general rule, the roofs of earthquake safe structures must be as light as possible. Many builders prefer profiled steel cladding on light-gauge steel Zed purlins or a double-skin with insulation and spacers.

References

- Halpin, Daniel W.; Senior, Bolivar A. (2010), Construction Management (4 ed.), Hoboken, NJ: John Wiley & Sons, p. 9, ISBN 9780470447239, retrieved May 16, 2015

- Main-construction-types: civilengineerblog.com, Retrieved 11 July 2018

- Advantages-of-structural-steel-frame-construction-407580-7: whirlwindsteel.com, Retrieved 22 May 2018

- "Cazza to build world's first 3D printed skyscraper". Jochebed Menon, Construction Week Online, March 12, 2017. Retrieved July 17, 2017

- Main-construction-types: civilengineerblog.com, Retrieved 31 March 2018

- McIntyre M, Strischek D. (2005). Surety Bonding in Today's Construction Market: Changing Times for Contractors, Bankers, and Sureties. The RMA Journal

- Top-ten-health-and-safety-risks-in-construction: alcumusgroup.com, Retrieved 22 April 2018

- Chitkara, K. K. (1998), Construction Project Management, New Delhi: Tata McGraw-Hill Education, p. 4, ISBN 9780074620625, retrieved May 16, 2015

- How-3d-printing-transforms-the-construction-industry: mbadmb.com, Retrieved 29 May 2018

- "Dubai and Cazza Construction Technologies Announce Plans to Build World's First 3D Printed Skyscrape". Claire Scott, 3D Print. March 13, 2017. Retrieved July 17, 2017

- Types-structural-steel-framing-systems-18554: theconstructor.org, Retrieved 31 March 2018

4

Foundation Engineering

Foundation is an essential aspect of an architectural structure. It connects to the ground and allows the transference of the load of the structure to the ground. Foundations are either deep or shallow. Some of the important considerations of foundation engineering are underpinning, framing, building envelope, house raising and basement waterproofing, which have been extensively discussed in this chapter.

The foundations of the building transfer the weight of the building to the ground. While 'foundation' is a general word, normally, every building has a number of individual foundations. Most buildings have some kind of foundation structure directly below every major column, so as to transfer the column loads directly to the ground.

Since the weight of the building rests on the soil (or rock), engineers have to study the properties of the soil very carefully to ensure that it can carry the loads imposed by the building. It is common for engineers to determine the safe bearing capacity of the soil after such study. As the name suggests, this is the amount of weight per unit area the soil can bear. For example, the safe bearing capacity(SBC) at a location could be 20 T/m^2, or tonnes per square metre. This figure is the maximum the soil can bear, so an engineer will take pains to see that her design does not exceed this figure in any part of the building.

This capacity also changes at different depths of soil. In general, the deeper one digs, the greater the SBC, unless there are pockets of weak soil in the earth. To properly support a building, the soil must be very firm and strong. It is common for the soil near the surface of the earth to be loose and weak. If a building is rested on this soil, it will sink into the earth like a ship in water. Building contractors will usually dig until they reach very firm, strong, soil that cannot be dug up easily before constructing a foundation.

To study the properties of the soil before designing foundations, engineers will ask for a soil investigation to be done. A soil investigation engineer will drill a 4" or 6" hollow pipe into the ground, and will remove samples of the earth while doing so. He will then send these samples to a lab to find out the detailed properties of the soil at every depth. Soil is usually composed of *strata*, or different layers, each with its own set of properties. Drilling technology today makes it easy and economical to drill to great depths, easily several hundred metres or more, even in hard rock.

The soil investigation team will then prepare a *soil investigation report* that lists the engineering properties of the soil at regular intervals, say every 2 meters. Based on this deport, engineers designing the structure can decide at what depth of soil to provide the foundations, the type of foundations they should provide, and the size of the foundations.

Every once in a while, engineers will find fill at a site. This occurs when humans have previously dug up the earth there, and then filled it back in. This happens if a quarry was dug or a building built there previously. Since fill is loose and soft and cannot support weight, engineers will dig to a depth below that of the fill, where strong soil is found, and construct foundations there.

The study of soil, and its properties and behavior, is called *soil mechanics.*

Once the foundations have been built, the loose soil that has been excavated must be put back over and around the foundations. This is called backfilling. Backfilling must be done carefully, as the soil there must support the weight of the floor slab at ground level (called the first floor in the US). Backfilling is done by putting back the soil in horizontal layers about a foot thick, and then compacting the earth, or squeezing it under pressure in a wet condition. This causes the soil particles to be squished together and removes air voids, there by making the layers strong. Good backfilling also improves the performance of the foundations, as the earth holds them firmly in place, and weighs down on the foundations to anchor them in position.

The act of strengthening a foundation is called *underpinning.* It can also be called foundation repair. While it is difficult to conceive of how foundations (that are underground in an occupied building) can be repaired or strengthened, there are techniques that can achieve this. In a sense, these methods can be likened to surgery performed on a patient.

Many structures like dams, bridges, buildings, roads etc. are constructed by civil engineers to serve our various requirements. All these structures are above ground and are visible. These are called superstructures. Structures apply load on soil on which they rest. If superstructures are placed directly on the soil, the soil gets overstressed and will not be able to support them safely. To safely transfer the load of the structure on soil, some part of the structure is placed below the ground.

This part of the structure is called sub-structure. Sub-structure is usually called foundation. Thus structural elements that connect, bridges, buildings etc. to the ground are called foundations. Foundation of any structure is very important because the safety and reliability of structure depends upon foundation.

Load of a structure are transmitted from the superstructure to the sub-structure i.e., foundation by columns or walls. The foundation distributes the load to the soil in such a manner that the soil is able to withstand the load as shown in below figure.

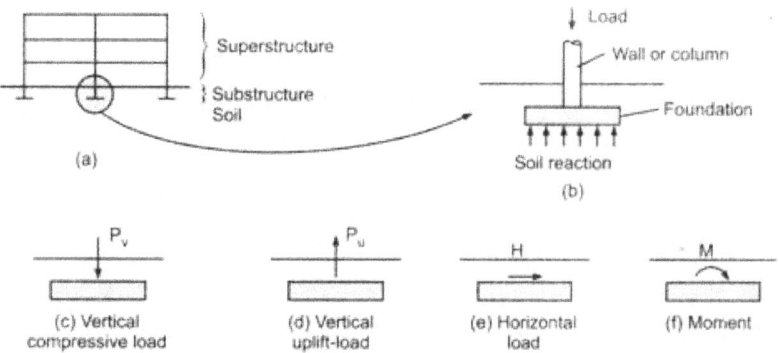

Load Transmission through the foundation

Types of Foundations

Foundations are Broadly Classified into two Categories:

- Shallow foundations and
- Deep foundations

Shallow Foundation

Shallow foundations are also called spread footings or open footings. The 'open' refers to the fact that the foundations are made by first excavating all the earth till the bottom of the footing, and then constructing the footing. During the early stages of work, the entire footing is visible to the eye, and is therefore called an open foundation. The idea is that each footing takes the concentrated load of the column and spreads it out over a large area, so that the actual weight on the soil does not exceed the safe bearing capacity of the soil.

Types of Shallow Foundations

Following are the types of shallow foundations:

1. Spread Footing
2. Combined Footing
3. Raft Foundation
4. Annular Slab or Ring Foundation.

1. Spread Footing

Foundation which spreads the load from a wall or column to a greater width is known as spread foundation or footing. The spread footing provided to the walls of a load bearing structure is known as wall footing, continuous footing, or strip footing. Spread footing may also be stepped footing as shown in figure. or tapered footing as shown in figure.

i. Strip Footings

Strip footing, shown in figure, is the first and most conventional footing used in the history of civil engineering and may be constructed of stone masonry or concrete. Strip footing constructed of stone masonry usually has a stepped cross section, similar to the one shown in below figure.

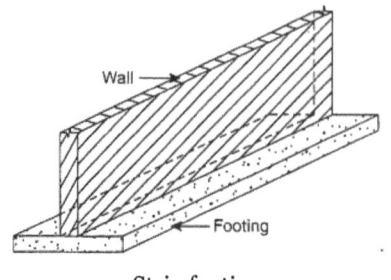

Strip footing

With the development of concrete, however, both load bearing structures and strip footings have become more or less obsolete, except for small lightly loaded residential buildings. It is also known as continuous footing or wall footing.

ii. Isolated Footing

Spread footing provided to the columns of a framed structure is called isolated footing, column footing, or pad foundation. Square column foundations, shown in figure, are the most economical but space restrictions between adjacent columns in a specific direction may warrant rectangular column footings, shown in figure. Circular footing, shown in figure, is not common and may be used for circular columns as the construction of form work and concreting may be more difficult for them than for square or rectangular footings.

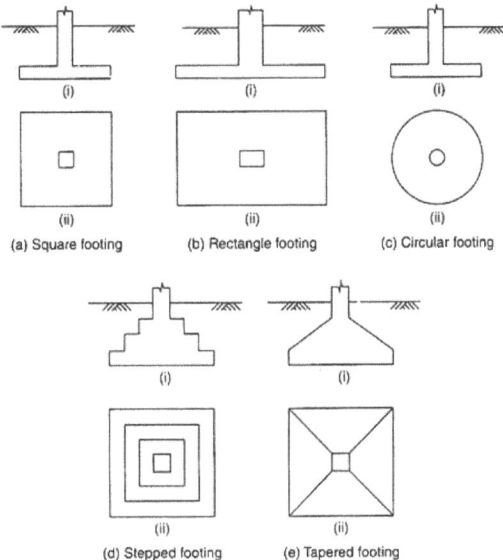

(a) Square footing   (b) Rectangle footing   (c) Circular footing

(d) Stepped footing   (e) Tapered footing

Types of isolated footing: (i) Section and (ii) plan

2. Combined Footing

Combined footing is used when footings of two adjacent columns are too close or overlap.

i. Rectangular Footing

Combined footing is most commonly rectangular, as shown in figure for equal column loads.

ii. Trapezoidal Footing

For unequal column loads, trapezoidal footing, shown in figure (b), may be used to ensure that the centre of gravity (CG) of the column loads coincides with the CG of the foundation in plan.

iii. Strap Footing

For footings situated near property lines, a strap footing, as shown in figure, is used to ensure that the edge of the footing near property line does not extend into the adjacent site. In this case, the footing near the property line (exterior footing) is connected to the footing inside the site (interior footing) through a strap beam. The strap beam transfers the load of the exterior column footing partially to the interior footing through structural action.

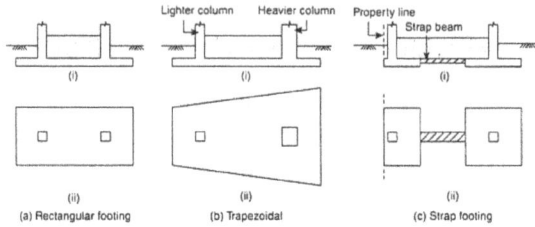

Types of combined footings: (i) Section and (ii) plan

3. Raft Foundation

Raft foundation, covering the entire area of the loaded structure with a slab, is provided when the total area of all footings is more than 50% of the loaded area. A raft foundation also called as mat foundation is also provided for heavy structures located over highly compressible and weak soils extending to large depth. The plan and section of a raft foundation is shown in figure.

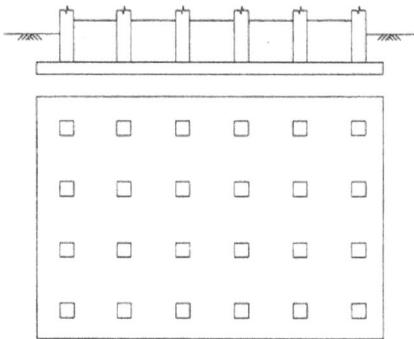

Raft foundation

4. Annular Slab or Ring Foundation

A ring foundation is sometimes provided for a large water tank with its columns connected through a ring beam and supported over an annular slab, as shown in figure.

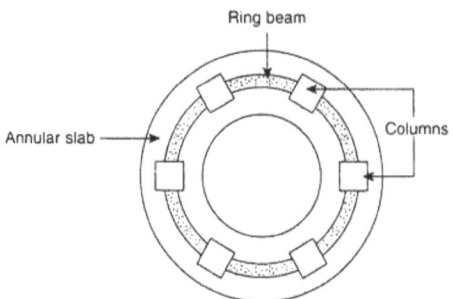

Annular slab or ring foundation

If a good bearing stratum of soil or rock is available at a shallow depth, a shallow foundation is the first choice as it is more economical and is always preferred to a deep foundation.

Deep Foundation

Deep foundation is required to carry loads from a structure through weak compressible soils or fills on to stronger and less compressible soils or rocks at depth, or for functional reasons. Deep foundations are founded too deeply below the finished ground surface for their base bearing capacity to be affected by surface conditions, this is usually at depths >3 m below finished ground level.

Deep foundation can be used to transfer the loading to a deeper, more competent strata at depth if unsuitable soils are present near the surface.

Types of Deep Foundation

The types of deep foundations in general use are as follows:

1. Basements

2. Buoyancy rafts (hollow box foundations)

3. Caissons

4. Cylinders

5. Shaft foundations

6. Pile foundations

1. Basement

These are hollow substructures designed to provide working or storage space below ground level. The structural design is governed by their functional requirements rather than from considerations of the most efficient method of resisting external earth and hydrostatic pressures. They are constructed in place in open excavations.

2. Buoyancy Rafts (Hollow Box Foundations)

Buoyancy rafts are hollow substructures designed to provide a buoyant or semi-buoyant substructure beneath which the net loading on the soil is reduced to the desired low intensity. Buoyancy rafts can be designed to be sunk as caissons, they can also be constructed in place in open excavations.

3. Caissons

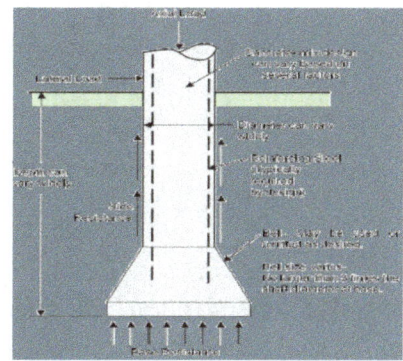

Caissons are hollow substructures designed to be constructed on or near the surface and then sunk as a single unit to their required level.

4. Cylinders

Cylinders are small single-cell caissons.

5. Drilled Shaft Foundations

Shaft foundations are constructed within deep excavations supported by lining constructed in place and subsequently filled with concrete or other pre-fabricated load-bearing units.

6. Pile Foundations

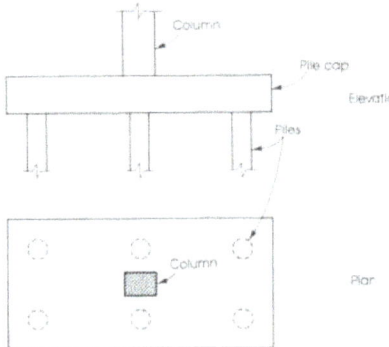

Pile foundations are relatively long and slender members constructed by driving preformed units to the desired founding level, or by driving or drilling-in tubes to the required depth – the tubes being filled with concrete before or during withdrawal or by drilling unlined or wholly or partly lined boreholes which are then filled with concrete.

Underpinning

Underpinning is a method for repair and strengthening of building foundations. Underpinning methods, procedures and their applications in strengthening of different types of foundations.

There are situations where a failure in foundation or footing happens unexpectedly after the completion of whole structure (both sub and superstructure). Under such emergency situation, a remedial method has to be suggested to regain the structural stability.

The method of underpinning help to strengthen the foundation of an existing building or any other infrastructure. These involve installation of permanent or temporary support to an already held foundation so that additional depth and bearing capacity is achieved.

Selection of Underpinning Methods

Underpinning methods are selected based on age of structure and types of works involved.

Structure Categories Based on its age:

- Ancient Structures :Age greater than 150 years

- Recent Structures : Age between 50 – 150 years

- Modern Structure : Age less than 50 years

Types of Works for Selection of Underpinning Methods

Conversion Works

The structure has to be converted to another function, which requires stronger foundation compared to existing

Protection Works

The following problems of a building has to undergo protection works:

- The existing foundation is not strong or stable

- Nearby excavation would affect the soil that supports existing footing.

- Stabilization of the foundation soil to resist against natural calamities

- Requirement of basement below an already existing structure

Remedial Works

- Mistakes in initial foundation design caused subsidence of the structure

- Work on present structure than building a new one

Structural Conditions which Requires Underpinning

There are many reasons that make an engineer to suggest underpinning method for stabilization of the substructure such as:

- The degradation of timber piles used as a foundation for normal buildings would cause settlement. This degradation of structures is due to water table fluctuations.

- Rise and lowering of the water table can cause a decrease of bearing capacity of soil making the structure to settle.

- Structures that are built over soil with a bearing capacity not suitable for the structure would cause settlement.

Need for Underpinning

The decision of underpinning requirement can be made based on observations. When an already existing structures start to show certain change through settlement or any kind of distress, it is necessary to establish vertical level readings as well as at the offset level, on a timely basis. The time period depends upon the how severe is the settlement.

Now, before the excavation for a new project, professionals have to closely examine and determine the soil capability to resist the structure that is coming over it. Based on that report the need for underpinning is decided. Sometimes such test would avoid underpinning to be done after the whole structure is constructed.

Methods of Underpinning

Following are the different underpinning methods used for foundation strengthening:

- Mass concrete underpinning method (pit method)
- Underpinning by cantilever needle beam method
- Pier and beam underpinning method
- Mini piled underpinning
- Pile method of underpinning
- Pre-test method of underpinning

Whatever be the types of underpinning method selected for strengthening the foundation, all of them follow a similar idea of extending the existing foundation either lengthwise or breadthwise and to be laid over a stronger soil stratum. This enables distribution of load over a greater area.

Different underpinning methods are mentioned briefly in the following topic. The choice of method depends on the ground conditions and the required foundation depth.

1. Mass Concrete Underpinning Method (Pit Method)

The mass concrete (or 'traditional') method of underpinning is an established technique, suitable for relatively shallow depths of underpinning. The method is often used for partial underpinning of sections of a building. This is probably the most common form of underpinning undertaken for residential properties. The method can be used in cohesive or granular soils, but is widely used in shrinkable clays. It can be used to prevent movement due to subsidence and heave, and is often used to assist in retro-fit basement construction.

Technique

The technique involves the construction of a new foundation beneath a failing section of a building by extending the existing footings down to a greater depth where stable soil of suitable bearing capacity exists. This is achieved by excavating individual bases in short lengths (usually not exceeding 1200mm) in a pre-determined sequence to a designed depth in a suitable stratum. The depth of the bases may vary depending upon the profile of the stratum selected. Once each base is excavated to the appropriate depth, and before concreting, supervisory staff and local authority officers inspect the excavation to check that the correct stratum has been reached and that the ground is free from soft spots, tree roots etc. Once the excavation has been approved, shutters are set in position and the base is backfilled with concrete of a specified mix. The concrete is usually cast to leave a narrow gap between the top of the base and the underside of the footing. When the concrete has cured sufficiently to reduce initial shrinkage and to support the applied load, a sand and cement 'dry-pack' is rammed into this gap to the transfer the building load to the new foundation. Alternatively under some circumstances, the concrete may be flooded up to the underside of the existing footing and well vibrated to ensure that any trapped air pockets are removed.

Base construction is repeated sequentially until the whole length of wall where underpinning is required has been supported. Bases will generally be linked together using 'joggle' joints to provide a key between adjacent bases. Reinforcement cages can be introduced using couplers to provide continuity between the bases. When specifying reinforcement for underpinning bases, it is important to consider the health and safety implications for the operatives. Anti-heave precautions consisting of polythene sheeting and/or low density polystyrene are usually installed when underpinning is constructed in shrinkable clay.

Advantages and Disadvantages

The main advantages of mass concrete underpinning are:

- Bases are usually constructed from one side of a wall only, and it is therefore often possible to construct all underpinning from outside without disturbing the inside of the building and possibly necessitating a building to be vacated.

- Soil conditions can be examined at close quarters, tested for strength using hand penetrometers or vane testers, and the presence of tree roots or soft spots easily identified.

- The method is not technically complex and operatives can be relatively easily trained to achieve competence.

- Excavations can often be undertaken using minimal amounts of plant and machinery.

- Acts as a root barrier which may help prevent damage to other parts of the building.

- It can be designed to act as a retaining structure to assist in retro-fit basement construction.

- By increasing the base width, it is possible to reduce imposed stresses by spreading the load in weak soil

- Disadvantages of this system include:

- There are large amounts of excavated material to be disposed of.

- There are large amounts of concrete to be imported to construct the bases.

- Excavations and bases are difficult to construct in unstable or water-logged ground.

- Base depths in excess of 3.0 metres are generally uneconomic and create a number of health and safety issues.

- Mass concrete underpinning generally requires good site access due to the amount of spoil to be removed and concrete imported. If access is difficult, the technique is more difficult and may prove costly.

2. Underpinning by Cantilever Needle Beam Method

This method is an extension of above-described pit method and is often named as Cantilever pit method of underpinning. This method is suitable if the building has a strong interior column and that the deepening of foundation can be done only in one direction due to site limitations.

In this method an underpinning pit is excavated beneath the existing exterior wall whose foundation is faulty. A needle beam is strapped to join the exterior wall with that of existing interior loaded column or pedestal. The exterior wall is hanged with the needle beam arrangement pushed up by a hydraulic jack supported on Fulcrum.

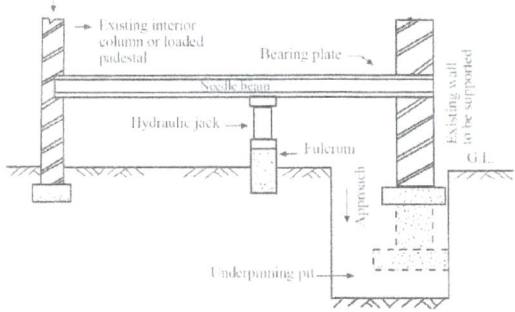

Figure: Extended Pit Method of Underpinning

This method is faster than the traditional method. With this procedure being adopted the load carrying capacity of foundation is increased.

Another slight modification of this method also exists in which the existing foundation is supported by a Cantilever beam strapped to tension piles and compression piles which were rested on firm strata underneath as shown in the figure below.

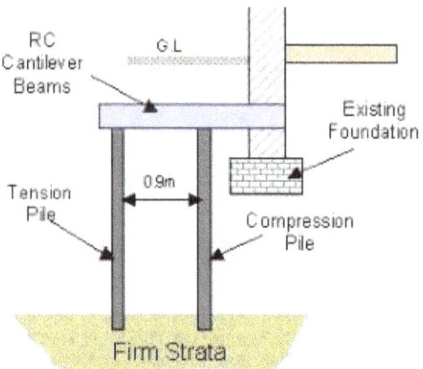

Figure: Cantilever Needle Beam Method of Underpinning-Sectional View

Advantages of Cantilever Needle Beam Method

- Faster than traditional method

- One side access only

- High load carrying capability

Disadvantages

- Digging found uneconomical when existing foundation is deep

- Constraint in access restricts the use of needle beams

3. Pier and Beam Underpinning Method

This method is developed to cover the limitation of mass concreting method. This method is good for relatively deeper foundation and is feasible for all the soil conditions. In this method reinforced concrete beams are placed supported on the mass concrete underneath the ground or piers as shown in figure.

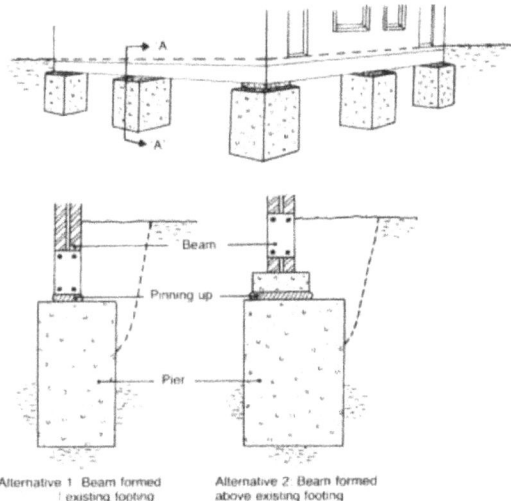

Figure: Pier and Beam Method

The ground conditions and the extent of load to be carried will decide the size and the dimensions of the piles. For foundation that is deeper than 6m, this method is economical.

Causes of Pier and Beam Foundation Problems

- Poor Drainage: Insufficient drainage can leave the foundation more prone to cracking and the growth of mold. Drainage can cause unwanted moisture when directed toward the basement, which can be caused by misdirected garden and lawn watering.

- Soil Problems: Moisture retention within the soil can cause sheetrock cracking, an unevenness in the basement floor, and seepage into the basement area through the foundation walls.

- Original Construction Problems: Pressure over time can cause old cedar piers to rot through. If rotting occurs, sinking can follow and cause severe degradation to the foundation.

Once the cause(s) of pier and beam foundation deterioration have been determined, a solution can be selected and implemented.

Primary Solutions to Pier and Beam Foundation Problems

- Foundation Shoring – Reinforcing the foundation with timber piers.

- Foundation Strengthening – Supporting the foundation with concrete columns.

Pier and Beam Advantages:

- Easy repair of utilities and wiring

- Ability to add extra insulation simply

- Less expensive to repair

Pier and Beam Disadvantages:

- Longer construction duration

- Greater susceptibility to bugs and rodents

- Expensive flooring

- Likelihood of frozen pipes and drainage problems

4. Mini Piled Underpinning

This method can be implemented where the loads from the foundation have to transferred to strata located at a distance greater than 5m. This method is adaptable for soil that has variable nature, access is restrictive and causes environmental pollution problems.

Of the various ways to strengthen a building's foundation, piling is the most versatile. Piles come in a wide range of types, but in recent years mini piles have increasingly become the most popular choice, especially for domestic properties.

Piling

Piling is used for projects from building large structures to adding a basement to your home. In particular, it's become a popular choice for underpinning an existing building to counter problems like subsidence or heave.

The two main ways of inserting piles are driving, where the pile displaces the soil rather than removing it first, and boring, where you create a hole for concrete or grout to be poured into. The biggest advantage of piling, which makes it suitable for everything from underpinning to building bridges, is that the piles can go right through weaker soil, transferring the load to the more robust soil or bedrock below.

Mini Piling

Mini piling is a variation on this, using piles with a narrow diameter. This makes them light and inexpensive, but still able to support a considerable load.

For the most common type of mini piling, a hollow steel shaft is either screwed or drilled into the ground. Grout or concrete can then be poured in to form the pile, with the soil supported throughout by the steel shaft. This means that, unlike the traditional boring method, no extra supports are needed, even in weak soil.

Alternatively, especially if there's limited headroom, sectional auger mini piles can be used. This involves boring down and adding multiple flight sections as you go. Depending on the soil's stability, the sections can either be removed or left in place while pouring the concrete.

5. Pile Method of Underpinning

In this method, piles are driven on adjacent sides of the wall that supports the weak foundation. A needle or pin penetrates through the wall that is in turn connected to the piles as shown in below figure.

These needles behave like pile caps. Settlement in soil due to water clogging or clayey nature can be treated by this method

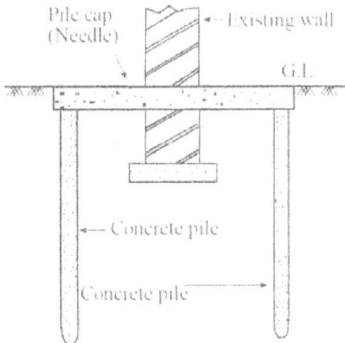

Figure: Underpinning by Pile Method

6. Pre-test Method of Underpinning

This method is suitable for buildings with 5 to 10 stories and the foundation type of pad and strip footing. The method will involve the initial compaction as well as compression of the underneath soil after excavation. Thus, soil is subjected to predetermined loads. All these procedures are done before underpinning. Raft foundation can also use this method. The pre-test method has less noise and disruption during construction.

Framing

Frame construction is a building technique which involves building a supportive framework of studs, joists, and rafters, and attaching everything else to this framework. This building style can be accomplished very quickly with a skilled crew, and it is extremely common all over the world. Most wooden homes, for example, are made with this style of construction, especially in the United States.

The process of frame construction starts with building up a sill on the ground, with the sill being attached to a foundation. Long studs are attached to the sill at set intervals to create a network which can be attached to the joists and rafters which make up the roof or additional stories. The frame may be additionally supported with cross bracing and other techniques. Essentially, frame-style construction creates a skeleton, and a fast crew can frame a house in only a few days.

Once the frame is complete, walls and other features can be added. The structure grows progressively more stable as stiff flooring and walls are added, creating additional support and resistance to the elements. Within the structure, the builders can differentiate from critical structural walls, which provide support to keep the building safe, and partitions which can be used to divide and change the shape of various spaces within the structure for utility.

Platform frame construction, in which a structure is built floor by floor, is the most common type of this construction style. Some older buildings utilize balloon frame construction, in which long joists run all the way from the sill to the top plate, which meets the roof, no matter how tall the building is. For practical reasons, balloon frame construction is usually limited to two to three floors, and it is uncommon to see in new structures, because of timber availability issues.

Classically, frame construction is accomplished with wood, which needs to be carefully cut and handled to ensure that the integrity of the frame is maintained. Wood which has not been cured properly, for example, will develop warping and twisting which could pull the structure out of alignment. Metal beams can also be used in framing, and they can cut costs significantly in areas where timber is expensive.

There are some issues with frame construction which need to be addressed carefully by builders. One of the biggest problems is that the spaces between the joists and the walls can be ideal conduits for fire, allowing fire to leap quickly from floor to floor. This type of construction is also vulnerable to rot and other types of damage. Although the studs are designed to be redundant so that the structure can stand if one fails, the failure of multiple neighboring joists can be catastrophic.

Walls

Wall framing in house construction includes the vertical and horizontal members of exterior walls and interior partitions, both of bearing walls and non-bearing walls. These *stick members*, referred to as studs, wall plates and lintels (sometimes called *headers*), serve as a nailing base for all covering material and support the upper floor platforms, which provide the lateral strength along a wall. The platforms may be the boxed structure of a ceiling and roof, or the ceiling and floor joists of the story above. In the building trades, the technique is variously referred to as *stick and frame*, *stick and platform*, or *stick and box*, as the sticks (studs) give the structure its vertical support, and the box-shaped floor sections with joists contained within length-long post and lintels (more commonly called *headers*), support the weight of whatever is above, including the next wall up and the roof above the top story. The platform also provides the lateral support against wind and holds the stick walls true and square. Any lower platform supports the weight of the platforms and walls above the level of its component headers and joists.

Framing lumber is subject to regulated standards that require a grade-stamp, and a moisture content not exceeding 19%.

There are three historically common methods of framing a house:

- Post and beam, which is now used predominantly in barn construction.

- Balloon framing using a technique suspending floors from the walls was common until the late 1940s, but since that time, platform framing has become the predominant form of house construction.

- Platform framing often forms wall sections horizontally on the sub-floor prior to erection, easing positioning of studs and increasing accuracy while cutting the necessary manpower. The top and bottom plates are end-nailed to each stud with two nails at least 3.25 in (83 mm) in length (*16d* or *16 penny* nails). Studs are at least doubled (creating posts) at openings, the jack stud being cut to receive the lintels(headers) that are placed and end-nailed through the outer studs.

Wall sheathing, usually a plywood or other laminate, is usually applied to the framing prior to erection, thus eliminating the need to scaffold, and again increasing speed and cutting manpower needs and expenses. Some types of exterior sheathing, such as asphalt-impregnated fiberboard, plywood, oriented strand board and waferboard, will provide adequate bracing to resist lateral loads and keep the wall square. (Construction codes in most jurisdictions require a stiff plywood sheathing.) Others, such as rigid glass-fiber, asphalt-coated fiberboard, polystyrene or polyurethane board, will not. In this latter case, the wall should be reinforced with a diagonal wood or metal bracing inset into the studs. In jurisdictions subject to strong wind storms (hurricane countries, tornado alleys) local codes or state law will generally require both the diagonal wind braces and the stiff exterior sheathing regardless of the type and kind of outer weather resistant coverings.

Corners

A multiple-stud post made up of at least three studs, or the equivalent, is generally used at exterior corners and intersections to secure a good tie between adjoining walls, and to provide nailing support for interior finishes and exterior sheathing. Corners and intersections, however, must be framed with at least two studs.

Nailing support for the edges of the ceiling is required at the junction of the wall and ceiling where partitions run parallel to the ceiling joists. This material is commonly referred to as *dead wood* or backing.

Exterior wall Studs

Wall framing in house construction includes the vertical and horizontal members of exterior walls and interior partitions. These members, referred to as studs, wall plates and lintels, serve as a nailing base for all covering material and support the upper floors, ceiling and roof.

Exterior wall studs are the vertical members to which the wall sheathing and cladding are attached. They are supported on a bottom plate or foundation sill and in turn support the top plate. Studs usually consist of 1.5 by 3.5 inches (38 mm × 89 mm) or 1.5 in × 5.5 in (38 mm ×

140 mm) lumber and are commonly spaced at 16 in (410 mm) on center. This spacing may be changed to 12 or 24 in (300 or 610 mm) on center depending on the load and the limitations imposed by the type and thickness of the wall covering used. Wider 1.5 in × 5.5 in (38 mm × 140 mm) studs may be used to provide space for more insulation. Insulation beyond that which can be accommodated within a 3.5 in (89 mm) stud space can also be provided by other means, such as rigid or semi-rigid insulation or batts between 1.5 in × 1.5 in (38 mm × 38 mm) horizontal furring strips, or rigid or semi-rigid insulation sheathing to the outside of the studs. The studs are attached to horizontal top and bottom wall plates of 1.5 in (38 mm) lumber that are the same width as the studs.

Interior Partitions

Interior partitions supporting floor, ceiling or roof loads are called loadbearing walls; others are called non-loadbearing or simply partitions. Interior loadbearing walls are framed in the same way as exterior walls. Studs are usually 1.5 in × 3.5 in (38 mm × 89 mm) lumber spaced at 16 in (410 mm) on center. This spacing may be changed to 12 or 24 in (300 or 610 mm) depending on the loads supported and the type and thickness of the wall finish used.

Partitions can be built with 1.5 in × 2.5 in (38 mm × 64 mm) or 1.5 in × 3.5 in (38 mm × 89 mm) studs spaced at 16 or 24 in (410 or 610 mm) on center depending on the type and thickness of the wall finish used. Where a partition does not contain a swinging door, 1.5 in × 3.5 in (38 mm × 89 mm) studs at 16 in (410 mm) on center are sometimes used with the wide face of the stud parallel to the wall. This is usually done only for partitions enclosing clothes closets or cupboards to save space. Since there is no vertical load to be supported by partitions, single studs may be used at door openings. The top of the opening may be bridged with a single piece of 1.5 in (38 mm) lumber the same width as the studs. These members provide a nailing support for wall finish, door frames and trim.

Lintels (Headers)

Lintels (or, headers) are the horizontal members placed over window, door and other openings to carry loads to the adjoining studs. Lintels are usually constructed of two pieces of 2 in (nominal) (38 mm) lumber separated with spacers to the width of the studs and nailed together to form a single unit. The preferable spacer material is rigid insulation. The depth of a lintel is determined by the width of the opening and vertical loads supported.

Wall Sections

The complete wall sections are then raised and put in place, temporary braces added and the bottom plates nailed through the subfloor to the floor framing members. The braces should have their larger dimension on the vertical and should permit adjustment of the vertical position of the wall.

Once the assembled sections are plumbed, they are nailed together at the corners and intersections. A strip of polyethylene is often placed between the interior walls and the exterior wall, and above the first top plate of interior walls before the second top plate is applied to attain continuity of the air barrier when polyethylene is serving this function.

A second top plate, with joints offset at least one stud space away from the joints in the plate beneath, is then added. This second top plate usually laps the first plate at the corners and partition intersections and, when nailed in place, provides an additional tie to the framed walls. Where the second top plate does not lap the plate immediately underneath at corner and partition intersections, these may be tied with 0.036 in (0.91 mm) galvanized steel plates at least 3 in (76 mm) wide and 6 in (150 mm) long, nailed with at least three 2.5 in (64 mm) nails to each wall.

Light wood framed construction is one of the most popular types of building methods for homes in the United States and parts of Europe.

It has the following characteristics:

- It is light, and allows quick construction with no heavy tools or equipment. Every component can easily be carried by hand - a house essentially becomes a large carpentry job. The main tool is a handheld nail gun.

- It is able to adapt itself to any geometric shape, and can be clad with a variety of materials.

- There are a huge variety of products and systems tailored to this type of construction.

It has these negative characteristics:

- It is not highly fireproof, as it is made of wood.

- It is not strong enough to resist major wind events such as tornadoes and hurricanes.

Every timber frame home structure is made of a few basic components:

Studs are vertical wooden members within the walls. Joists are the horizontal wooden beams that support the floors. Rafters are the sloping wooden beams that support the roof.

Sheathing are the sheets that are nailed over the studs to connect them securely and form the wall surfaces. Siding is the exterior cladding that covers the walls from the outside.

Let us examine the major types of light wood framed structures.

Balloon Frame Structures

While this is an outmoded form of wood construction no longer used today, it is good to know what it is and why it is no longer used.

In balloon frame construction, if you had a two-storey house that was twenty feet high, you would use a single 20 foot long vertical stud for both storeys. This made the studs heavy and difficult to handle. The second problem was the gap between the two studs, which acted passageways for the spread of fire from the lower to the upper storey.

For these reasons, balloon frame construction has been superseded by platform frame construction, which is superior in all respects.

Platform Frame Structures

This is the sequence you would follow to erect a 2-storey platform frame house.

- Erect the ground floor platform, a horizontal wooden platform over the foundation.

- Build all the walls upto a height of one storey. This can easily be done by building each wall flat - on the floor platform - and then tilting it vertically.

- Build the next floor platform.

- Erect the next set of vertical walls in the same manner on the top of the second platform.

- Build the sloping roof over the walls.

As you can see, this system uses shorter, lighter studs that are easy to handle. It is much easier to build walls flat and then tilt them into place. Since each floor is a separate horizontal platform, this makes it convenient for construction workers to move around. These platforms also break the vertical spread of fire.

The only disadvantage of platform frame vs. balloon frame construction is that wood shrinkage plays a bigger role in platform frames.

Building Envelope

One term that's used a lot when building a house is 'building envelope' or 'building enclosure.

The concept of a building envelope relates to design and construction of the exterior of the house. A good building envelope involves using exterior wall materials and designs that are climate-appropriate, structurally sound and aesthetically pleasing. These three elements are the key factors in constructing your building envelope. The building envelope of a house consists of its roof, sub floor, exterior doors, windows and of course the exterior walls.

A tight building envelope is preferable in cooler climates.

Tight and Loose Building Envelopes

A building envelope is normally referred to as either 'tight' or 'loose'. A loose envelope allows air to flow more freely through the building, whereas a tight envelope restricts air or controls how it is admitted. Australia's climate (as varied as it is), is such that a tight envelope is generally the preferred choice. Innovations in the design and materials of exterior walls increasingly allow us to take advantage of the environment and use the outside of the house to regulate the climate indoors.

Environmental separator is another term used to describe the enclosure, but note that this generic term also applies to separators of two different interior environments. The term building enclosure is preferred to the term building envelope largely because it is considered both more general and more precise. Also note that the building enclosure may contain, but is not the same as, the so-called thermal envelope, a term that is used to refer to the thermal insulation within the enclosure. The enclosure, the loadings it must resist, and its functions are addressed in this digest.

The Nature of the Building Enclosure

Both the above-grade and the below-grade portions of the building enclosure are part of a physical system involving three interactive components: the exterior environment(s), the enclosure system, and the interior environment(s). The exterior environment above grade is very different from that below grade, and within any building there can be numerous interior environments. The nature of the building enclosure and its spatial relationship with the other parts of the building are shown in Figure.

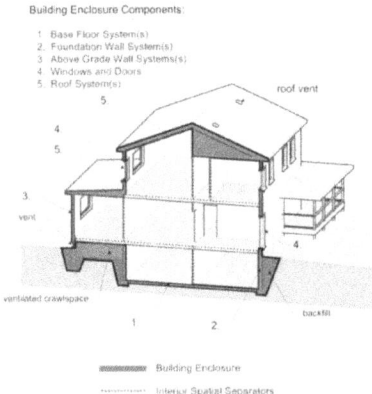

Figure: The building as a set of separated spaces and the as-built separators

As suggested by figure, a building in general consists of a collection of spaces bounded by a set of spatial separators. There are separators between interior environments as well as separators of an interior environment and the exterior environment; collectively, the latter constitute the building enclosure. In many respects, the functions of a separator of internal environments within a building (floors and interior walls) are very similar to those of the building enclosure. Usually, however, the performance requirements of internal separators are fewer and much less onerous.

Enclosure Components

The primary function of the enclosure is to separate the interior environment from the exterior environment to which it is exposed. Physically, the typical building enclosure usually consists of the following components:

- The roof system(s)

- The above-grade wall system(s) including windows (fenestration) and doors

- The below-grade wall system(s), and

- The base floor system(s).

The building enclosure should not be thought of as a combination of numerous one-dimensional or even two-dimensional planar components. Each enclosure component is a three-dimensional, multi-layer, multi-material assembly that extends from the inside face of the innermost interior layer (e.g., the paint or wallpaper) to the outside face of the outermost layer (e.g., paint or roof shingles). The overall enclosure is made up of all the contiguous enclosure sub-assemblies.

Each enclosure component is an assemblage of layers of specified products (such as gypsum board or wallpaper) or materials (such as paint or wood). For instance, the thermal insulation could consist of a layer of blown-in glass fiber filaments. A deliberate air space or cavity is also a considered to be a layer. Consider, for example, the sloped roof system shown in Figure; this constitutes an enclosure assembly consisting of all the contiguous layers between the finish on the ceiling and the outer face of the roof shingles. Furthermore, the compacted gravel under the basement floor (or base floor) system and the backfill to the outside of the foundation or basement wall system are each specified layers within their as–built assemblies. As such, they can be considered to be an integral part of their respective building enclosure components. By common definition each enclosure sub-assembly incorporates all the contiguous (in-contact) layers.

The climate-related loadings that a building and its enclosure actually experience are modified versions of the local climate. These microclimates are modified by adjacent buildings, the landscaping (especially trees), and other parts of the enclosure. Roof overhangs, for instance, are climate modifiers to the walls below as well as integral parts of the roof system.

The above definition of the enclosure has some limitations. First, the physical difference between site microclimate modifiers and the enclosure is not that straightforward. Second, although one can define the enclosure as the building component that separates interior from exterior environments, it is not always easy to precisely define "interior" and "exterior." For example, the backfill around a basement wall or ivy growing on a masonry wall are, by definition, considered to be part of the enclosure. They are also part of the site. What cannot be ignored is the significant impact of this soil or ivy on moisture and thermal loadings.

Consider also an attached garage. The space in the garage does not experience outdoor conditions, nor is it conditioned like the interior of the building. The enclosure could be considered to incorporate both the exterior wall of the house and the garage wall, with the space between (containing the car) being an integral part of the enclosure. Alternatively and preferably, the enclosure could be considered to be the garage wall, and the intermediate wall could be viewed as an interior separator with significantly different conditions on each side.

Similarly, two adjacent internal spaces can also have quite different environmental conditions. For example, consider the not uncommon scenario of a hotel with a warm humid swimming pool next to an air-conditioned exercise room or a community center containing both a swimming pool and ice arena. In such cases the loadings and performance of the interior separator are more like those of a building enclosure component. In this type of condition enclosure design principles should be applied to the interior separator.

The influence of climate modifiers, such as trees, garages, overhangs, and the soil on the loadings experienced by the enclosure is not always appreciated. Especially in the case of low-rise buildings, the effect on the enclosure from the exterior environment can be intentionally reduced or moderated by the use of microclimate modifiers such as plantings, fountains, building overhangs and windbreaks. For example, a roof overhang (an integral part of the roofing system) will moderate the amount of rain on the wall below and will therefore be a microclimate moderator for the vertical portion of the building enclosure. With taller buildings it is more difficult to modify the external microclimate.

The definition of where the enclosure begins and exterior environment stops can sometimes be confusing, especially in the case of buffer spaces such as garages, screened porches, vented and storage attics, or vented crawlspaces.

The exterior environment could be considered to be a three-dimensional space with randomly varying mass and energy properties. The local climate and weather provide the major, but not the only, exterior loadings. Local climatic records – for instance, average and extreme values – are commonly available but these data are usually based on open-field, weather-station records. The building may be some distance from the weather station, and the terrain (hills, other buildings, etc.), the landscaping (trees, shrubs, etc.) and the building itself (overhangs, protrusions, etc.) can moderate the weather that each enclosure component is actually subjected to. As a matter of convention, that portion of the exterior environment that is close to, and affected by, the building is called the external microclimate. In fact, different parts of the enclosure are subjected to different exterior microclimates.

The interior environment is usually specified in relation to the physical needs of people and are defined in terms of temperature, relative humidity, airflow rate, and air quality. However, sensitive equipment, materials, or processes (e.g., computers, archived paper, electronics factories) may require different, often more stringent, conditioning requirements than those required by (adaptable) human occupants. Special uses, such as swimming pools, pharmaceutical plants, ice arenas, etc have significantly different conditions than normal. Because the interior spaces are usually conditioned, the state of the interior environment is often assumed to be constant, but in practice, significant variations occur over the day, between different spaces and over the seasons.

The service systems (e.g., lights, motors, stoves) and the contents of a space have a strong and time-dependent influence on the interior environment. The level of the influence depends on the controls, output capacity, and distribution effectiveness.

Figures indicates that, while the building enclosure separates the interior and exterior environments, it really experiences several microclimates. The enclosure interacts with both environments and, in turn, affects both environments. This interaction is usually dependent on the time of day, the day of the week, and the season. There is thus a cyclical (diurnal, workday/non-workday, weekly and seasonal) aspect to the time-dependent response of the enclosure as it separates and, to some degree, modifies the influence of both environments.

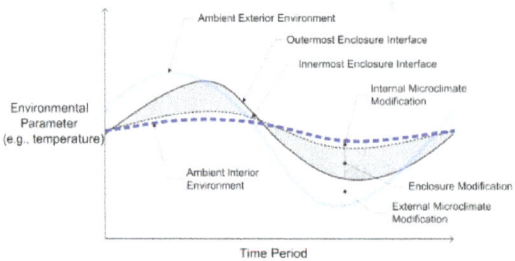

Figure: Contributors to environmental modification

While the enclosure may be the largest modifier, the difference or the modification between the ambient and interface conditions can also be significant. Note that the scale of Figure will be different for different parameters. Moreover, this simple representation of variation over one cycle of time for one parameter does not accurately or fully represent the complex behavior and interactions that occur.

The properties of exterior environments and interior environments are separate and major fields of study. Each environment is discussed in other Building Science Digests, Primers, and Design Guides.

Enclosure Loadings

The performance of any portion of the enclosure should be considered in relation to the various loadings generated by i) the exterior environment, ii) the interior environments, and iii) the enclosure itself. The generic term loading refers to any event, phenomenon or characteristic that can affect the enclosure. Each and every loading that impacts the building enclosure can be categorized into one of just five types, namely:

- Gravity related
- Ground related
- Heat related (thermal)
- Moisture related
- Air related

Exterior Environment Loadings

The exterior environment described above is responsible for many different loadings. Sources of these loadings are the climate (weather over the short term), human or human-made effects, and nature or natural phenomena. Table identifies, by type and source, those loadings due to the exterior environment that could affect the enclosure. It should be evident from this list of loadings that the design or analysis of the enclosure involves consideration of not only the usual structural loadings but also mass (air, moisture, etc.) and energy (heat, light, sound, etc.) loadings. Moreover, for design purposes, we need to have some knowledge of both average and extreme conditions for each relevant loading.

Type / Source	Heat related	Moisture related	Air related	Ground related	Gravity related
Weather or natural climate	Ambient conditions, solar	RH, fog, rain, ice, snow	Barometric pressure, wind		Water, snow, hail
Abnormal climatic effects	Reflected solar, lightning	Tornado, hurricane, flooding	Tornado, hurricane	Frost heave, landslide	Wind-borne missile
Natural phenomena	Fire, Ground water	Adfreezing, Freezing	Radon, methane, soil gas	Seismic, land- slide, settlement, termites, plants, etc.	Hydrostatic pressure, soil pressure
Human-made weather	Global warming, city effect (2-7 ^{0}C)	Smog, Acid rain	Wind related vortex/swirl		
Human-induced events	Fire	Fire (hoses, sprinklers, etc.)	Smoke, sonic boom, sound, explosion		Impact, wear and tear

Table: Loadings from the exterior environment

Interior Environment Loadings

The interior environment consists of occupied, used and, often, conditioned spaces. As shown in Table, the main sources of loading are human-induced, the result of natural phenomena or due to conditioning. Variations in the state of the interior "climate" and the related extreme values tend to be smaller or less than those occurring within either the enclosure or the exterior. This occurs not only because the building enclosure modifies the effect of exterior variations but also because the interior environment is generally augmented by inputs of mass and energy, e.g., the building or parts of it may be mechanically conditioned.

For a building to successfully meet its desired performance, the required state of each of the interior environments must be ensured. The state of an interior space may either be maintained constant or may vary in time as a function of the nature and degree of space conditioning, internal gains (equipment, lighting, solar effects and human activities) and the modifying influence of the enclosure.

Type / Source	Heat related	Moisture related	Air-flow related	Ground related	Gravity related
Interior space	Ambient conditions, solar	RH, water (sprinklers, etc.)	Barometric pressure, wind, stack, fan induced		Water
Natural phenomena	Fire	Fungal growth, mold	Radon, methane	Settlement, termites, plants, etc.	
Human-induced events	Fire, people	People, flooding, combustion, equipment	Smoke, sound, explosion		Impact, wear and tear, dead & live loads

Table: Loadings from the interior environment

Loadings from the Enclosure

The enclosure itself is a source of loading and, as shown in below Table, loadings arise from the enclosure element under consideration as well as adjacent elements.

To properly design buildings and their enclosures it is necessary to have some knowledge and understanding of the relevant environments as well as their interdependence. A considerable body of empirical and experimental data exists on the interaction of climate, site and building. Research, notably that by Fanger and Givoni, has identified and defined the physiological aspects of human response to both interior and exterior climates, and these data have been distilled into convenient forms for building designers

The enclosure is only one part of the larger system that moderates or controls the environments on both sides (especially the inside) of the enclosure at some location on the building. The building itself and the enclosure both influence the exterior microclimate that interfaces with the exterior surface of the building enclosure at a specific location. Consider, for example, the local impact of wind, rain, and solar radiation. In addition, the mass and energy inputs provided by heating, ventilation, air conditioning equipment or other internal gains likewise modify the interior environment.

Type Source	Heat related	Moisture related	Air related	Ground related	Gravity related
Element or component being considered	Volume change, shape change, fire	RH, built-in moisture, volume change, fungal growth, mold, creep, shrinkage, etc.	Off-gassing, air flow, air pressure differentials		Self weight, live loads
Adjacent Elements	Volume and shape change, fire	Volume change	Smoke		Dead loads, live loads

Table: Loadings from the enclosure

The Functions of the Enclosure

At the most basic level, the primary function of the building enclosure is to separate the interior and exterior environments. In practise the building enclosure has to provide the "skin" to the building, i.e., not just separation but also the visible façade. Unlike the superstructure or the service systems of buildings, the enclosure is seen and is therefore of critical importance to owners, the architect and the public. The users or occupants are concerned with both sides of the building enclosure. The appearance and the operation of the enclosure have an influence on the interior environment and on factors such as productivity and satisfaction.

In general the physical function of separation of the building enclosure may be grouped into three sub-categories, as follows:

1. Support functions, i.e., to support, resist, transfer and otherwise accommodate all the structural forms of loading imposed by the interior and exterior environments, by the enclosure, and by the building itself. The enclosure or portions of it can be an integral part of the building superstructure – usually by design but sometimes not.

2. Control functions, i.e., to control, regulate and/or moderate all the loadings due to the

separation of the interior and exterior environments – largely the flow of mass (air, moisture, etc.) and energy (heat, sound, etc.).

3. Finish functions, i.e., to finish the enclosure surfaces–the interfaces of the envelope with the interior and exterior environments. Each of the two interfaces must meet the relevant visual, esthetic, wear and tear and other performance requirements.

A fourth building-related category of functions can also be imposed on the enclosure, namely:

4. Distribute functions, i.e., to distribute services or utilities such as power, communication, water in its various forms, gas, and conditioned air, to, from, and within the enclosure itself.

Figure illustrates a representative portion of a building enclosure, its functions and the nature of the loadings. The enclosure may also serve other, usually non-physical, purposes such as advertising or as a symbol or image, e.g., as a statement about power or security or wealth, etc., but these are not considered here.

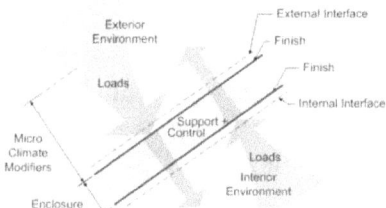

Figure: The enclosure and its functions

Each and every part of the enclosure must satisfy the relevant support, control, finish and distribution functions. Only the support and control functions are needed everywhere. The finish functions may not be needed in some areas (such as above suspended ceilings, in service rooms). Moreover the distribution functions, which largely service the adjacent interior spaces, need to be met only where there is a service or utility to be distributed – large segments of most enclosures do not need to fulfill this function.

	Specific loadings	Interior	Support	Control	Exterior
Essentially structural	Gravity – Dead (assembly, etc.)		●		
	Gravity – Live (people, snow, etc.)		●		
	Wind		●	○	
	Ground Movement (seismic, settlement, etc.)		●		
	Explosion		●		
	Rheological (creep, shrinkage, etc.)		●		○
	Impact (vehicles, missiles, people, etc.)		●		
	Fire		●		
Essentially environmental	Heat (thermal, etc.)	○		●	
	Air (pressure, movement, leakage, etc.)	○		●	
	Moisture (built-in, precipitation, etc)	○		●	○
	Smoke			●	
	Solar radiation (incident, reflected, etc.)			●	○
	Chemical attack/atmospheric (acid rain, etc.)			●	○
	Particulate matter (dust, VOC's, etc.)			●	
	People (wear & tear, etc.)	○		●	
	Insects, birds, animals, (termites, rodents, etc.)			●	
Essentially perceptual	Light (natural, incandescent, fluorescent, etc.)			●	
	Sound	○		●	○
	Visual – local	●			●
	Visual – contextual	●			●

(Left-hand spanning label: **Causal phenomenon or loading**)

Primary significance ●

Secondary significance ○

Tertiary significance ·

Table: General category of loadings and related functions

To further demonstrate the relationship between category of function and loading, consider Table. All the possibly relevant loadings are listed and their functional relevance is identified. This table is important because it clearly shows that the number of loadings is considerable and that each loading affects at least one functional sub-category for the enclosure and often more than one.

Basement Waterproofing

The basement waterproofing systems prevent water from leaking into the basement and damaging the foundation and wood. Basement waterproofing systems are needed whenever a basement, cellar, or other room is built at ground level or below. They are particularly important in areas where it is likely that ground water will raise the water table.

Without proper basement waterproofing systems, the basement can become cracked, and water pressure can result in serious damage to walls, as well as the growth mold and wood rot. Building the right kind of waterproofing system depends very much upon the environment in which the house is built.

Interior Sealant

Using interior sealants as basement waterproofing systems is a temporary measure, to be used during winter months to prevent snow and frost from raising the water table.

These sealants are often found in chemical spray form, and can be directly applied to walls and floors. Interior sealants can also prevent humidity and condensation within the basement. They can be absorbed by woods and porous building materials, causing cracks in masonry, damage to concrete, and wood rot. Masonry is also protected against spalling.

The most common use of these sealants is to prevent humidity inside the house from affecting the walls of the basement. For more secure measures against ground water, you will need to have either an exterior sealant or a drainage system.

Exterior Sealant

Exterior basement waterproofing systems stop ground water from reaching the basement walls as well as prevent mold and other damage which can occur in wet basement areas.

Waterproofing an exterior is the recognized IBC method to prevent damage caused by water. Exterior sealants were once only asphalt-based damp proofing, but now the most common kind of exterior basement waterproofing systems use a polymer base. This kind of material will last for the life of the building.

Other exterior sealants will probably erode in 20 years. You will not usually have to add anything to an exterior sealant, which should applied during the construction of the house, particularly in areas where flooding, heavy rain, or hurricanes are likely to occur.

Water Drainage

Drainage can be used to mitigate basement water and is often considered to be another form of basement waterproofing. Water drainage functions by drawing water away from the foundation, and forcing it into a drain, or through a pump system. Pump kids are available which can be installed in DIY form or by professional plumbers.

Water drainage basement waterproofing systems often need to be run on an isolated electric system in case of power shortage, especially during periods of storms. Sump pumps should be placed in a pit and sealed with a lid in order to keep the water away from the electricity. Doing so also prevents humidity in the pump from entering the atmosphere of the basement. Keeping the lid airtight ensures that poisonous gasses won't seep into the house.

House Raising

Raising a house (a.k.a house elevating and house jacking) is the process of house raising and building a new, or extending the existing, foundation below it. During the elevation (house raising) process most houses are separated from their foundations, raised with hydraulic jacks, and held by temporary supports while a new or extended foundation is constructed.

There are many reasons people choose to raise their homes. Some of them the most common reasons is flood protection, to reduce flood insurance rates and to add more space to their home.

House Raising: Reasons to Raise your House

If you were a Hurricane Victim

If your house was flooded by Hurricane Sandy or Irene and you'd like to prevent it from happening again, you should consider house raising and raise your house above the Base Flood Elevation. The BFE is the computed elevation to which floodwater is anticipated to rise during a base flood and is a regulatory requirement for the elevation or flood proofing of structures. Once a structure is raised, only the foundation remains exposed to flooding while the living area is safe from damage.

If your Home is in a Flood Zone

Whether or not you were a victim of the storm, if your home is in a flood zone, you should raise your house to protect it from future flooding. These zones are shown on flood maps. Moderate to low risk areas are zones that begin with the letters "B," "C" or "X" and high-risk zones start with the letters "A" or "V."

The most vulnerable homes are in the "V" zones, which are waterfront areas that are at highest risk for flooding and subject to 3-foot breaking waves. But houses still need to be raised in the coastal "A" zones, which isn't as vulnerable as the "V" zone but is still subject to major damage.

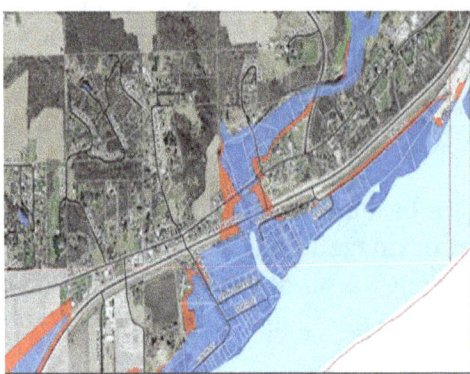

Even houses in "B," "C," or "X" zones should consider house raising. Roughly 25% of all flood insurance claims are filed in low to moderate flood-risk areas.

To Lower your Flood Insurance

House raising will also lower your flood insurance in the long-run. Under the Flood Insurance Reform Act of 2012, you could save more than $90,000 over 10 years if you raise your house 3 feet above the Base Flood Elevation.

To add More Space

House raising also adds more usable space under your home. It's the perfect alternative if you're looking to add more square footage, but you don't want to build an extension that, for example, takes away backyard space. This added space could be used for storage, a garage, or a nice basement.

If you have a small yard to begin with or you'd like a covered area outside to hosts guests, house raising is a great option for that too. Instead of building continuous walls after the lift, you could build separate piers, posts, columns, or pilings to create an open space.

To Strengthen, Repair or Replace a Foundation

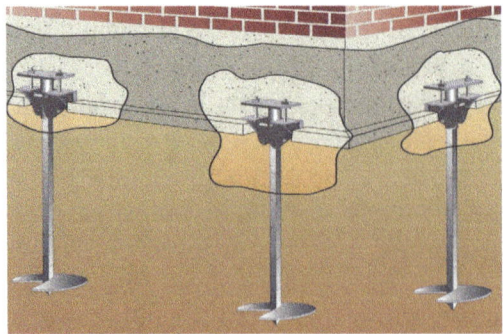

The most effective way to repair or replace a foundation is by first raising or raising and moving your house off of the existing one.

If a foundation needs to be strengthened or repaired, it may go through a process called underpinning, it may need helical piles or it may need both. Underpinning, to put it simply, is when a foundation is installed under the existing one. A lot of foundation issues occur due to poor soil. If this is the case for your home, a type of steel called helical piles is drilled into the foundation and far below the surface into bedrock. This is considered a long term or permanent fix to foundation issues caused by ground or soil movement.

If the damage to your foundation is serious and widespread enough, then it will have to be replaced. The current foundation is excavated to create room for demolition and construction. After the debris have been cleared, new footings are poured, forms are set for pouring foundation walls, the concrete is cured and helical piles are installed if necessary.

Once the foundation has been strengthened, repaired or replaced, the house is then lowered or moved back and lowered to attached it to the new, stable and sturdy foundation.

To Increase the Value of your Home

House raising increases its value and potential resale by an estimated 15% to 25%. Adding square footage alone increases its value, the more space, the more it's worth. As for potential resale: houses in flood zones that aren't raised are nearly impossible to sell because flood insurance rates are so high.

Buyers would have to purchase additional flood insurance on top of buying your house, which would cost them thousands of dollars a year as seen above in figure. This may discourage them from buying your home and encourage them to buy one that is already raised. Elevating a home may be expensive, but it would be worth the cost if you are having trouble selling it.

Methodology

The methodology adopted to achieve above objectives Comprise of following steps.

1. Surveying of the residential building is to be done before starting the process of house lifting. It's important to study the weak members and the members of the building which requires the support before lift.

2. Load calculation of the building is done to get the numbers of jacks to apply for lifting the weight of the building and the numbers of jacks are applied according to the area of the

building. Total load of the building is divided with the capacity of jacks which gives the number of jacks required to lift the building.

Load calculation for the building is as below:

Dead load of slab = 0.10*1*25 = 2.5 KN/m² Floor finished = 10.0 KN/m²

Live load = 3.0 KN/ m² Total load = 6.5 KN/ m²

Factored load = 1.5*6.5 = 9.75KN/m²

Load transfer from slab to beam = 9.75 KN/m² Load calculation for beam

Self weight of beam = 0.23*0.45*25 = 2.58 KN/m² Parapet wall load on beam = 2 KN/m²

Total load transfer = 14.33 KN/m³

Wall load of super structure = 0.23*22.55*20 = 11.73 KN/m²

Total load coming to the foundation = 11.73+11.73+14.33+14.33+2.58 = 54.7 KN/m²

Load in ton= 5.49/m²

$$\text{Jacks requirement per m}^2 = \frac{\text{Total load}}{\text{Capacity of one jack}}$$

$$= 5.49/2$$

$$= 2.75 \text{ jack/ m}^2 \text{ required}$$

Therefore, 3 jacks required per m²

3. Before starting the lifting of the building it's important and necessary to disconnect the supplies of the building like electricity, gas connection, drainage connection etc. to avoid the interruption in the work and for the safety of the people working.

4. The support to weak members is provided to avoid the falling of members down during the process of lifting as the safety precautions to ensure the safety of the building and workers. Generally the supports of I- beams are provided as the supports.

5. At First the excavation is done near the walls for the application of the jacks, the jacks are applied below the ground beam or supports of steel beams.

6. The jacks are applied in the space of the excavation and the jacks are applied and the house is lifted by jacking the jacks simultaneously. The jacks are removed and the parallel brick masonry is done to support the lifting of the building.

7. The brick masonry is to be done to act as the foundation of the building, it supports the whole building and this ultimately increases the height of the building.

8. The jacks are removed after the brick masonry is completed and can sustain or bear the load of the building.

9. The pebbles and murrum are filled in the plinth area of the building. The backfilling of the sand should be well compacted to support the floor load of the building.

10. The flooring is done after the compacted soil filling. After the completion of the flooring the supply connections are connected.

11. After the lifting of the house is done the filling of cracks is done with the cement grouting is done

Estimation

The demolition cost of the building is excluded for the estimation of new construction because the scrap value is assumed equal to demolition charge and therefore the cost of demolition is excluded.

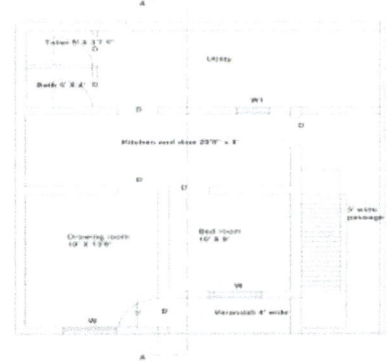

Figure : Ground floor plan of residential building

Estimation of the house lifting method by jacks includes cost for brick masonry work, flooring, plinth filling and base material filling. The cost also includes charges of filling cracks etc. if any found during the lifting of the house.

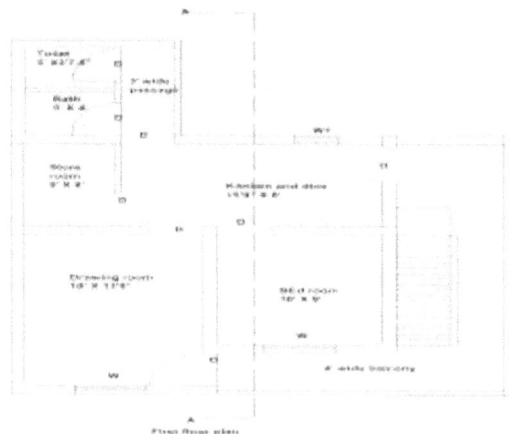

Figure : First floor plan of residential building

1. Estimate for New Construction

Total quantity of different items for Ground floor and first floor for new construction is combined and presented in table.

Item No.	Particulars of items and details of work	Quantity
1	Earthwork includes excavation in foundation	71.64m^3
2	B.B.C.C of proportion (1:4:8) with brick bats including compaction curing etc.	16.53 m^3
3	First class brick masonry in foundation CM (1:6)Up to plinth including curing etc.	32.28m^3
4	Back filling with ordinary soil in foundation trenches	26.66m^3
5	Back filling with yellow soil in plinth including watering, compaction and levelling etc.	43.7m^3
6	R.C.C work in ground beams including formwork, compaction curing etc.but excluding reinforcement.	2.98m^3
7	First class brick masonry in super structure including scaffolding, curing etc. in Cement mortar (1:6) 0.23 m thick. up to bottom level of beam. After deduction for door and window	47.10m^3
8	First class brick masonry in W compound wall and Bath and toilet 0.23 thickness in Cement mortar (1:6)	06.84 m^3
9	First class Brick masonry in cement mortar (1:6) including curing in compound wall, bath room, toilet, and stair way railing	36.93m^2
10	RCC work including formwork, scaffolding, compaction, curing etc.Excluding reinforcement in lintel, weather shed, beam and slab	21.45m^3
11	Plastering and pointing 12 mm thick plastering with 1:6 cement local sand mortar in walls.Including scaffolding	534.95m^2
12	Flooring 2.5 cm Cement concrete (1:2:4) floor	42.56m^2

13	Flooring With ceramic tiles Inside the house	86.24 m^2
14	Dado with wall tiles	28.76m^2
15	Walls tile - Cream Royale/ Riviera in kitchen	09.22 m^2
16	Wood work in Door and windowSal wood work in chaukhats in door and window	00.26m^3
17	Sal wood work in chaukhats in doorand window, 3 cm thick paneled shutters of Deodar, wood in door and window	24.40 m^2
18	Doors and windows fittings of oxidized iron	24.40 m^2
19	Steel and Iron work Steel reinforcement TMT bars including bending in RCC work@ 1% of RCC work Excluding steps	19.28 MT
20	Iron work in hold fast and window bars	396.24 kg
21	Painting two coats over one coat of Priming for door and window	48.72m^2
22	Distempering one coat after one Priming coat	396.5 m^2
23	Parapet wall -0.10 m thick First class brick masonry in cement mortar (1:6) Including curing, on terrace.	30.18 m^2
24	Plastering and Pointing - 12 mm thick CM 1:6 Both side of parapet.	33.50m^2
25	Water proofing- Using china mosaic and 50 mm thick Cement mortar 1:6 including finish-ing, levelling and curing etc.	55.65 m^2

2. *Estimate for House lifting*

Total quantity of different items for house lifting is presented in table.

Table: Quantities of different work items for House Lifting

Item No.	Particulars of items and details of work	Quantity
1	Excavation for placing of jacks from plinth to the depth of footing	9.40 m^3
2	First class brick masonry above foundation up to 1.5 m	12.97 m^3
3	Backfilling of rubble and sand murrum including compaction with watering and levelling	64.91 m3
4	Flooring with ceramic tiles on ground floor above compacted soil backfilling.	43.27 m^2
5	Filling of cracks by cement grouting including finishing	1 job
6	Jack application	1 job

References

- Maass, John (1957). The Gingerbread Age – A View of Victorian American. New York City: Crown Publishers. p. 140. ISBN 0-517-01965-5

- Foundation-engineering-meaning-and-types-of-foundation-45683: yourarticlelibrary.com, Retrieved 25 March 2018

- Indiana DNR, Division of Histo; ric Preservation and Archaeology. "Historic Building Research Handbook" (PDF). Retrieved June 13, 2013

- Underpinning-methods-procedure-applications-14480: theconstructor.org, Retrieved 16 April 2018

- McKeever, D.B.; Phelps, R.B. (1994). "Wood products used in new single-family house construction: 1950 to 1992" (PDF). Forest Products Journal. Retrieved March 3,2007

- What-is-underpinning-methods-and: iamcivilengineer.com, Retrieved 11 March 2018

- Woodward, George Evertson (1865). Woodward's Country Homes. New York City: Geo. E. Woodward. pp. 151–152. ISBN 1-112-22157-3

- Wood-framed-construction: understandconstruction.com, Retrieved 20 July 2018

- Cavanagh, Ted (1999). "Who Invented Your House?". Who invented your house (text only) | Ted Cavanagh - Academia.edu. American Heritage of Invention and Technology Magazine. Retrieved February 23, 2016

- 3-types-of-basement-waterproofing-systems-explained: doityourself.com, Retrieved 29 May 2018

- Kosny, J.; Desjarlais, A.O. (1994). "Influence of Architectural Details on the Overall Thermal Performance of Residential Wall Systems". Journal of Building Physics. Retrieved March 3, 2007

5

Structural Systems in Building

In architecture, the term 'structural system' refers to the load-resisting sub-system of an architectural structure. The transference of loads occurs through interconnected elements. Some of the structural elements in buildings are walls, floor, roof, ceiling, etc., which have been adequately covered in this chapter. Some of the diverse topics in this chapter also address the important facets of ventilation and plumbing system.

Structural System

Structural system, in building construction, the particular method of assembling and constructing structural elements of a building so that they support and transmit applied loads safely to the ground without exceeding the allowable stresses in the members. Basic types of systems include bearing-wall, post-and-lintel, frame, membrane, and suspension. They fall into three major categories: low-rise, high-rise, and long-span. Systems for long-span buildings (column-free spaces of more than 100 feet, or 30 metres) include tension and compression systems (subject to bending) and funicular systems, which are shaped to experience either pure tension or pure compression. Bending structures include the girder and two-way grids and slabs. Funicular structures include cable structures, membrane structures, and vaults and domes.

Structural Systems and Materials

A structural system is defined as the main part of the building which carries and transfers the loads, both vertical and lateral, safely to the soil through the foundation. Structural systems consist of many elements, including: floor slabs, decks, joists, beams, girders, shear walls, columns, braces, and foundations. Structure systems involve horizontal and vertical systems. Horizontal systems include bending resistant systems (beam systems, Vierendeel, folded plate, cylindrical shell); axial resistant (truss, space truss, tree); form resistant (arch, vault, dome, grid shell, HP shell, etc.); tensile resistant (suspended and stayed systems, etc.). Vertical systems include: moment frame, braced frame, shear wall, cantilever, framed tube, braced tube, bumbled tube, suspended high-rise, etc. Each system has advantages as well as different purposes. For example, a moment frame is recommended for use in office buildings in order to provide free flexible space for rental purpose. Shear walls are good to use for hotels and apartments to provide lateral force resistance as well as sound transmission isolation and unit separation. Furthermore, each structural system has a different level of stiffness. For example, a moment frame is more flexible than a braced frame system. In contrast, a shear wall system is more rigid than a braced frame system. Shear walls and braced frames are better to resist wind load, while ductile moment frames are better to resist seismic load.

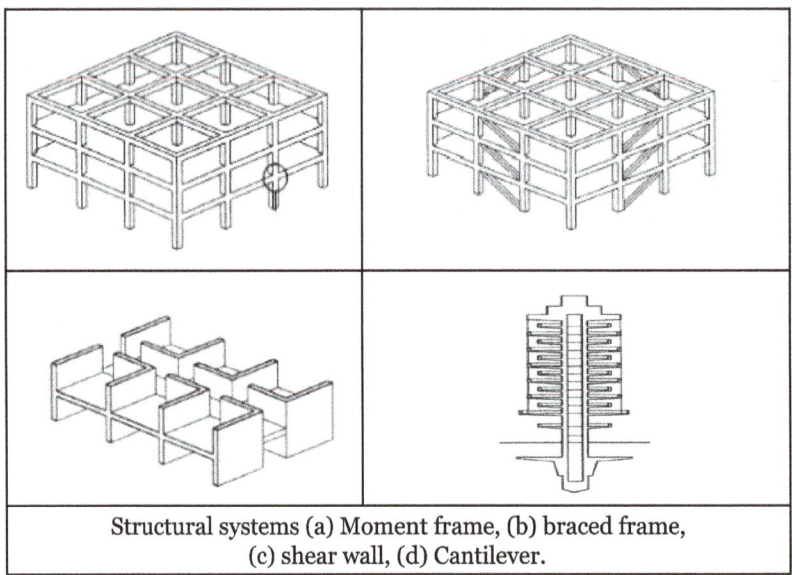

Structural systems (a) Moment frame, (b) braced frame,
(c) shear wall, (d) Cantilever.

There are also several alternative structural building materials: steel, concrete, masonry, wood, aluminum, and fabric. Each material has different properties as well as different purposes. For example, wood buildings are the most popular buildings in the United States because they are light, minimize seismic load, as well as wood is readily available and relatively cheap compared to concrete and steel. In contrast, concrete buildings are the most popular buildings in Saudi Arabia because wood is not available and concrete is good to resist wind load. Concrete is also cheaper than steel which is less available.

Selection of Structural Systems

The selection of structural system is sometimes based on personal experience or perception without being evaluated as it should be to provide advantage for the project. The conceptual selection process provides an orderly way to determine and review vital criteria which leads to the selection of the optimum structural system. The structural system should be integrated and accommodated to other building systems, like: architectural, mechanical, electrical, and building services.

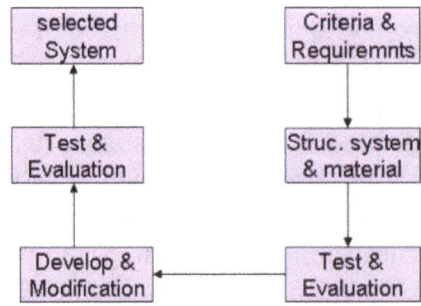

Selection of structural system process

However, the selection of the structural system is often passed through many processes, as shown in above figure. At the beginning of the process the criteria and requirements should be determined. The second step is applying different structural systems to the criteria and requirements.

The third step is testing and evaluating the performance of each structural system. The forth step is developing and modifying the tested system and retesting as well as evaluating it again. The last step is selecting the optimum structural system and material.

Selection Criteria for Structural System

The structural system is considered as the most vital system in the building for many reasons. First, it usually has the highest cost compared to other building systems cost. Second, the structural system is required for a building to stand up. The other building systems should be accommodated and adapted to it, for example, the mechanical equipment, air conditioning ducts and other services must be integrated with the structural system and elements. Finally, the structural system should save occupants and properties from natural forces, such as gravity, wind and seismic loads.

The selection criteria method is very wide in scope and includes not only structural aspects but also architectural considerations. Determining the required criteria depend on owner needs, project constraints, and project requirements.

Structural Criteria

Material Cost

Material cost affects the selection of structure system. Material cost depends on many factors like availability, energy consumption for production, abundance or shortage and economic condition of the country. Obviously, local material cost will be much cheaper than imported material. For example, in Saudi Arabia, the Hadeed Company which is considered the biggest provider of steel in the Middle East imports steel from Mexico and Brazil; which causes steel costs in Saudi Arabia to be more expensive than steel prices in the United States. On the other hand, concrete cost in Saudi Arabia is much cheaper than concrete cost in the United States because of the resources and the production abundance as well as the labor cost, notably for formwork.

Labor Cost

Labor cost depends on many factors, like location, the skill of labor, complexity of the work, as well as the economical situation of the country. However, the labor cost also is influenced by the same factors that affect the material cost. For example, in poor countries the labor cost is less than the labor cost in rich countries. In addition, the skilled labor cost is much higher than unskilled labor cost. For instance, the Saad Company, a construction company in Saudi Arabia, pays about $500 per month for a skilled mason while they pay about half as much for unskilled masons, i. e. $250 per month. Furthermore, the hourly labor rates differ from country to country. For example, the hourly rate for skilled mason in Saudi Arabia is $5 while the rate for a skilled bricklayer in the United States is $36.81.

Integration and Synergy

A structural system should be accommodated and integrated with other building systems. For example, as shown in figure, a structural floor slab, designed as one-way joist rib- slab, the gaps between the joists are used to pass the services, like air conditioning ducts, lighting, electric wires and water pipes. Furthermore, to avoid the obstacles of the main girder, all girders are designed to be inverted and hidden beam in order to ease the services passage under the beam. In addition, the structural module is designed in correlation with the architectural module, which provided the interior partition walls exactly over the slab joist in order to avoid slab shear over-stress.

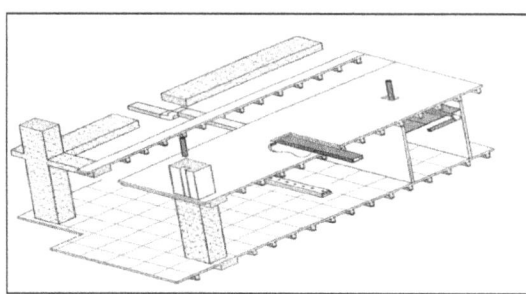

Building Systems Integration

Ease of Construction

Time means cost in construction. Therefore, engineers should select the fastest and easiest construction system to save the time and cost but provide quality control. The ease of construction depends on many factors like labor skills, design form and the type of materials. For example, the labor skills can control the time and quality of the work. Skilled labor can accomplish higher quality than unskilled labor. Also skilled labors need less time than unskilled which will reduce the project cost. In addition, the design form or shape will affect the ease of construction as well. For example, formwork for round columns costs less than square columns in the US.

Also, the type of material affects the ease of construction. For example, in the United States, steel usually costs less than concrete because it is prefabricated while concrete requires formworks and much time.

Span Limits

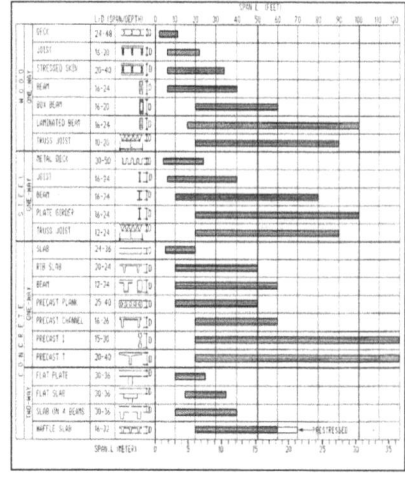

Span ranges for structure elements.

The span limit affects the stiffness of a system. For example, short spans are stiffer than long spans. In addition, different materials have different span limits as shown in below figures. For example, steel frames have longer span capacity than concrete frames. Hence, the span selection is affected by many factors like materials type, structural systems type, building type. There is a strong relationship between the span and the thickness of members. For instance, when the span increases the thickness will increase as well. Furthermore, increasing the span will increase the cost. On the other hand, minimizing the span much than the normal will increase the material quantity as well as the cost. Hence, the selection of span should moderate the balance of the long or short span. Figure shows span limits and span/depth ratios for structure elements and materials. Figure shows span limits and span/depth ratios for structure systems and material.

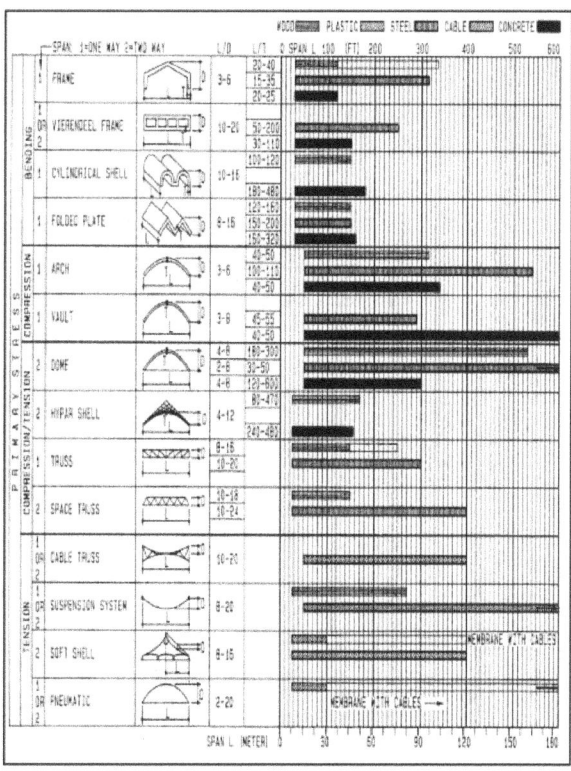

Span ranges for structure systems.

Gravity Load

Structural systems should be able to carry and resist vertical gravity load. There are many types of gravity load like dead load, and live load which includes snow load. Gravity load depends on many factors like building location, building occupancy, structural material and structural system. For example, in regions without snow load, like Saudi Arabia, the building structure should be designed for dead and nominal live load. In contrast, in mountain areas like Switzerland the building should be designed to resist actual snow load. Mountain snow load may be about 20 times greater than the nominal load in an area without snow load. Figure shows various types of gravity loads, like: Dead load (1), Live load (2), Distributed load (3), Uniform distributed load (4), and Concentrated load (5).

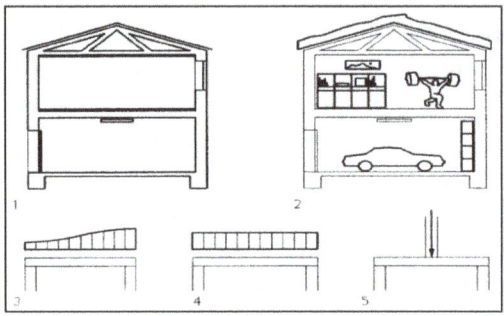

Gravity load types

Seismic Load

Earthquakes cause waves that transfer through underground layers until they reach to the ground surface. These waves affect building structures by shaking building foundations. Consequently, the building will respond to this motion. Seismic waves are defined by two terms as can be seen in figure. The first term is the *period* which is shown as horizontal line and defined as the time of one wave cycle and considered the most critical issue. For example, a building with a period that is resonant with the ground period could even collapse. The second term is the *amplitude* shown as vertical ordinate, defined as "the displacement of a wave perpendicular to the direction it moves".

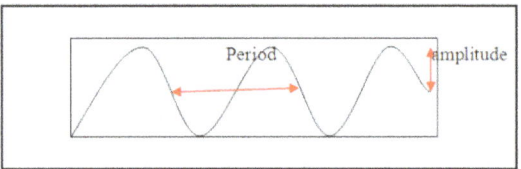

Figure : Seismic waves.

The earthquake's affect depends on many factors. For instance: the earthquake's magnitude which are measured by the Richter scale. The Richter scale created by Charles Richter in 1935 at the California Institute of Technology observes and measures the earthquake magnitude. Figure defines the Richter magnitude as follows: Left line plots earthquake distance, right line plots amplitude recorded on a seismograph; center line plots Richter Magnitude; defined by a line connecting distance to the amplitude. For example, magnitude 4.5 is minor earthquake, while 7.0 is a violent one.

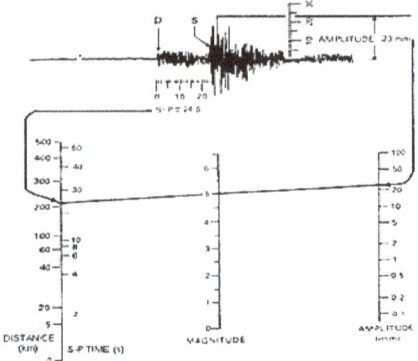

Richter scale method

There are several types of seismic waves. For example, body waves consist of *P waves* (Primary waves) that travel at high speed of 42,000 km/h), and *S waves* (Secondary waves) that vibrate normal to the wave direction and affect the building by dancing action. In addition, there are surface waves: Love waves and Raleigh waves.

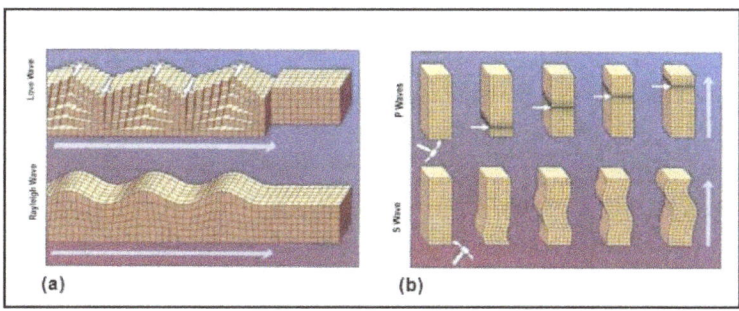

(a) Surface waves (b) Body waves

Seismic force affects the building as base shear, which basically follows Newton's law:

$$F = m \times a \text{ (force = mass} \times \text{acceleration)}$$

Due to this equation, structural engineers tend to minimize building mass and maximize ductility. Therefore, to minimize building weight, requires selecting lightweight materials and minimize structural members. Building design should resist seismic forces to protect the building, people and property. Most people think that rigid and strong buildings are best to resist seismic load. However, they are absolutely wrong: rigid buildings are subject to greater seismic forces than ductile ones.

Noise Control

The selected structural material should be able to isolate the noise in buildings. For example, in courthouses, hospitals, apartments and office buildings the selected material should prevent or reduce sound transmission in order to provide privacy. Concrete buildings have good sound rating according to standardized test. Because concrete has high mass and density, it has excellent sound isolation, sound absorption, and sound transmission reduction. For example, in apartment buildings, the apartments should be separated by party walls to reduce sound transmission. The party walls can also be shear walls and provide fire proofing. In addition, concrete structures can resist vibration and electrical interference especially the high-density.

Fire Safety

Fire safety is a critical issue to select the structure material. For example, concrete is considered the best material to resist fire, compared to steel which requires fire proofing to resist fire.

Steel starts to lose its strength at any temperature more than 300° C and begins to decrease strength at stable rate at about 800° C. The melting temperature of steel is about 1500° C (Colin Bailey). Hence, steel requires fire proofing (spray-on or other fire proofing).

Concrete has low thermal conductivity, 50 times lower than steel. In addition, it heats so slowly because of the density of aggregate and cement. Therefore concrete is ranked the best fire resistant material (Colin Bailey). Hence, concrete does not require fire protection like steel.

Sustainability (Durable and Recyclable)

Structural systems and materials should provide durability and sustainability to protect the environment. Concrete is considered a natural material and widely available. For example, the limestone, sand and clay, the main source of Portland cement production, are almost without limit. In addition, fly ash from coal burning, gravel, sand, and crushed stone are the main source of aggregate. Furthermore, the cement manufacturing process uses combustible waste and tires as a fuel source. Recycling of steel in the United States every year is more than recycling of aluminum, plastic and glass combined with the industry's overall. The recycling of steel rate is 64%. "Each year, steel recycling saves energy equivalent to powering about one-fifth of the households in the United States (about 10 Million homes) for one year" . Each ton of recycled steel saves about 1.400 pounds of coal, 2,500 pounds of iron, and 120 pounds of limestone.

Strength, Stiffness, Stability and Synergy

This criterion is considered a most significant issue to select the structure system and it depends on many factors like types of loads, height of building, span limit and material specification. As can be seen in the below figure, the structure system should satisfy:

- Strength to avoid and resist breaking.

- Stiffness to avoid extreme deformation.

- Stability to resist and avoid structural collapse.

- Synergy to support and integrate the architectural design.

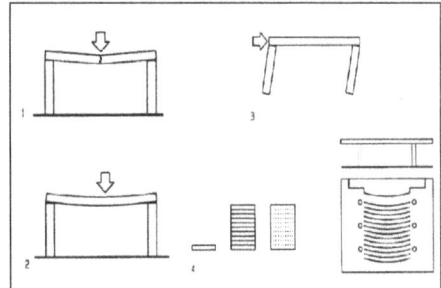

Structural members under vertical and lateral loads

Corrosion and Moisture Resistance

Corrosion resistance greatly depends on the type of material. Carbon dioxide in the air is the main reason for corrosion, and corrosion is causing reduction of the lifespan of structures. However, concrete ranks as the best material to resist corrosion and moisture compared to other materials because of its density and mass. For example, concrete protects steel bars from corrosion. Structural concrete members should be designed to meet the code requirements to provide corrosion protection for steel bars. Exposed steel may be protected by galvanizing, painting, or epoxy.

Material Transportation

Material should be selected locally in order to reduce transportation cost and to ensure getting it on time. Cast in place concrete requires less transportation than steel, which requires prefabrication in a shop while concrete is cast and mixed on site which requires only ready-mix transportation. Transportation costs depend on the distance between the project and factory. For instance, imported material will increase shipping costs and may causing delay in construction schedule which will affect the cost of project. In contrast, if the project is closed to the factory will be more economical and sustainable. Transportation will also control and limit element size. Hence, transportation affects the selection of structural materials.

Environmental Impact and Energy Consumption

Reduction of environmental pollution or impact is a new policy required, for a green building and healthy environment. Before selection of structural material one should have enough knowledge about the different materials and their environmental impacts. For example, one study investigated two different types of structural beam. One of them is reinforced concrete beam and the other is I- steel beam as shown in figure.

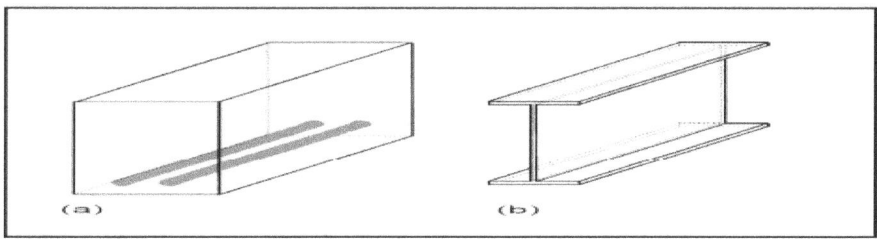

Schematic structures (a) concrete beam (b) I steel beam.

This study was done using the ATHENA database and computer program to study: Water pollution, air pollution, solid waste, energy consumption, resource use, and global warming potential. The investigation was done in Canada, which has the same cost and material availability as the United States. The test results show in table

Resource use of about 48.85 kg reinforcement concrete and about 18.69 kg of steel. Table also shows concrete and steel produced the same high mount of carbon dioxide, and produced the same amount of solid waste as well. Steel produced 3 times more water pollution than concrete. Steel also produced more air pollution than concrete. Finally, the energy consumption for steel is much higher than the energy consumption of concrete.

Hence, the production of concrete is better than steel because it is consuming less energy and less environmental impact .

Impact	Reinforced concrete	Steel
Resource use (kg)	48.85	18.69
Warming potential (kg equivalent CO_2)	9.97	8.95
Water pollution index	0.34	0.98
Air pollution index	2.01	2.46
Solid waste (kg)	1.87	1.80
Energy (MJ)	140.18	229.69

Environmental Impact of Reinforced Concrete and Steel Beams

Low Maintenance Cost

The initial cost is very important to select structural material but the running cost, or maintenance cost is also important. To estimate initial cost is relatively easy but to predict running cost is more difficult. Even though, the life span of concrete buildings is expected to be longer than that of steel buildings, whether steel and concrete are required periodic maintenance and inspection. For example, concrete should be isolated from water to prevent water penetration to reach the steel bars that causes corrosion. On the other hand, steel needs also painting, and maintaining the fireproofing and corrosion resistance, which causes steel failure.

Design Possibility

The shape and size of buildings and structural elements is one of the most vital issues to determine the type of structural material. Concrete can have any form since concrete seeks the form". In addition, concrete allows adding an extra floor or any horizontal extension as well. Furthermore, concrete could be switched from another type of material like steel. For example, the New York City developer built Hotel and Tower at the former Chicago Sun-Times site. This building switched from structural steel to concrete so that two additional stories could be added to the 1,125-foot building. Also, the concrete slab could be a flat slab which does not require beam space and thus will provide extra clear height for each floor. On the other hand, steel could be used for large span projects while concrete is recommended for short spans. Steel can provide extra clear height because the element size is much less than concrete.

Material Availability

Selection of a local structural material is an axiomatic issue because imported material will increase the cost because of shipping costs and limitation of transportation. However, the material availability differs from country to country. For example, in Saudi Arabia concrete is the most available material because the source of cement, aggregate, and sand are provided while steel considered as imported material from Mexico and Brazil. In contrast, the United States considered as one of the main producer of steel in the world. Even thought, steel cost has increased in the last two years America is still able to produce about 6 million tons per year. The total usage per year in the United States is about 4 million tons in the field of construction. Therefore, 2 million tons per year is considered as redundant. On the other hand, most of states had shortage of cement in 2004 because of event like the Florida hurricanes. Some of the United States have many companies who produce cement, and thus build most of building of concrete. Cement companies will increase by 2008 in these regions.

Building Type

Selection of structural system and material depends greatly on the building function. For example, for factories or large stores the best system is a long span steel frame, while short spans are good for housing and hospital buildings. Office buildings are better with moment frame or braced frame in order to provide flexibility for partitions. By contrast, apartment buildings require party walls to provide privacy, sound insulation, and fire resistance. Courthouses and hospitals require the best material for noise isolation.

Building Location

Building location also affects selection of structural system and material. For example, if the building is located in seismic zone, the building should be of lightweight material in order to minimizing building mass. In contrast, heavyweight material is bets in hurricane zones in order to maximize building mass to resist the wind load. Location by country will also affect the selection of materials. For example, in Saudi Arabia it is better to use concrete in order to resist the wind load and reduce cost while wood or steel are better in the United States because these resources are abundant. Furthermore, each country has its own building code, constrains, and policies which are different in each country.

Building Height Limit

This criterion is also a significant criteria to select structural systems and materials. Fazlur Khan defined optimal structure system for various building heights, defined by number of floors. Hence, building height depends on factors like structural system, structural material, and drift control.

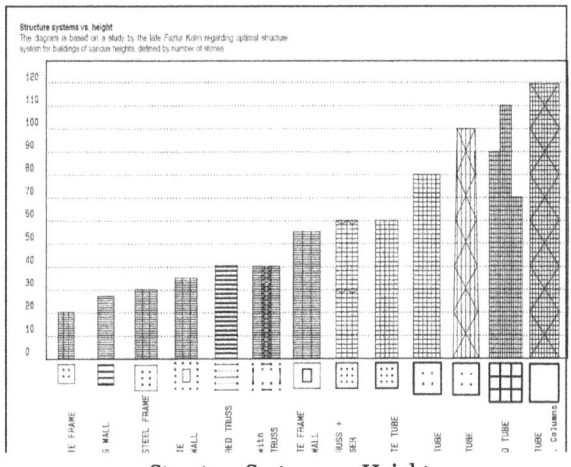

Structure Systems vs. Height

Code Requirements

Structural design must follow building codes used in the country. Structural materials must have the specification required by code. Building codes are different from country to country. There are many building codes, such as: International building code (IBC), uniform building code (UBC), Indian code, and Egyptian code, for example. In addition, some country has several codes. For instance, the United States used to have several codes: UBC in the west, BOCA on the east coast, and Southern Building Code in southern states. Furthermore, some countries are using the other's code; like in Saudi Arabia the current official used code is the International Building Code IBC 03. Building codes may be adapted with variations depending on weather condition, land topography, and other special conditions in seismic and hurricane zones. Building codes are updated about every three years, based on new research and conditions.

Site Condition (Access and Storage)

One constraint to select structural system and material is the site condition. The site conditions include the access to the site, storages area, site location, and the site topography. For example, if the site access is small or limited, that will limit the type of equipment and trucks that can reach the site. A site with large storage area could use cast in place concrete, while prefab systems are better for tight sites. Also, site location is important to minimize transportation distance and costs. For instance, for a project in Los Angeles downtown with traffic jams, the contractor should get authorization to determine a convenient time for material delivery.

Technology Availability

More and more, the field of architecture and structure has new developments and technologies. Further, technology of structural systems and material may be available in one country but not in other one. Hence, the selected system and material should be available locally. For example, some countries have no steel production but produce concrete for export In this case; use of the local material will minimize cost and avoid transportation costs. Technology includes equipment and tools. For instance, crane height available may control the maximum height of a building.

Energy Efficiency and Thermal Mass

The performance of structural material can be determined and measured by the energy efficiency of the selected material. Each material has its own heat properties, like heat absorption, transmission, and reflection. For example, concrete reduces temperature changes in summer and winter season as well as adobe and stone structures do. In addition, concrete releasing and absorbing heat causes reduction of energy consumption three ways:

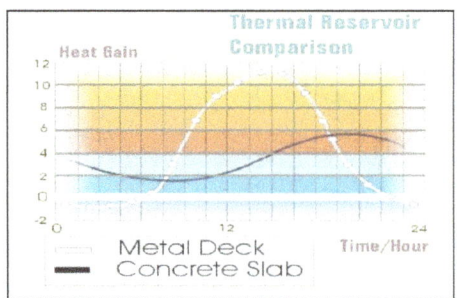

Thermal reservoir comparisons

First, concrete requires less energy to maintain constant interior temperature which saves the monthly bill. Second, concrete has greater time lag between cooling load and peak heating which delays heat transmission, consequently will reduce the monthly bill. Finally, concrete reduces initial building cost by using small cooling and heating equipment because the concrete thermal mass lowers the cooling and heating loads as can be seen in figure. the difference of heat gain between the concrete slab and steel metal deck through 24 hours.

Soil Class

There is a strong correlation between soil and building foundation Selection and design of building foundation should be done after analyzing and examining the soil. There are six types of soil in table: soft clay, stiff clay, sand compacted, gravel, sedimentary rock, and hard rock (granite). Each type has a different capacity as shown in the below table, ranging from 2 ksf (100 kPa) for soft clay to 200 ksf (9600 kPa) for the hard rock soil.

Soil type	Soil capacity (approximate)	
Soft clay	2 ksf	100 kPa
Stiff clay	4 ksf	200 kPa
Sand, compacted	6 ksf	300 kPa
Gravel	15 ksf	700 kPa
Sedimentary rock	50 ksf	2400 kPa
Hard rock (granite)	200 ksf	9600 kPa

Soil types and capacity

For example, if the soil is poor, it should be replaced with other type of soil or should be excavated to a strong layer of soil. However, rock type soil can be build on and need no excavation. On the other hand, some soft soils may require piles extending to stiffer layers as shown in below figure.

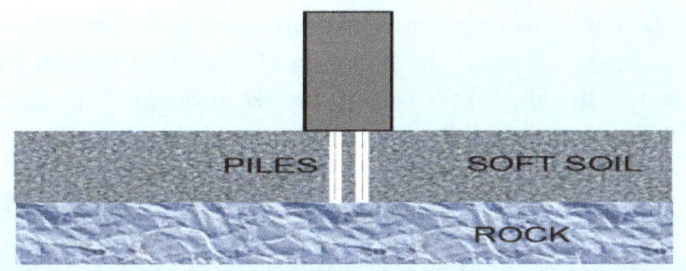

Figure: End bearing piles

Healthy Living and Indoor Air Quality

Indoor air quality affects healthy life, and requires appropriate specifications of material. Sick buildings are building with materials which produce gases or other organic material. Each material has its own percentage of pollution, but concrete has the lowest interior environment impact compared to other finish materials. Furthermore, concrete does not require fire proofing or other coating material which causes indoor air pollution.

Morphology

Correlation and integration of structure system with the building function should be a major objective. Each type of building has compatible structure systems and may not be compatible with others. For example, for office buildings, moment frame or braced frame systems provide flexible space for rental purpose; while apartments and hotels require party walls for sound insulation and privacy that can also be used as shear walls and for fire insulation. For exhibit halls and showrooms it is best to use large span structures to provide flexible space with few columns to avoid blocking of views. By contrast, apartments and hotels need no long span and may have short span to minimize structure depth and cost and provide better stiffness. Structures should also correlate with the architectural morphology. For example, trusses limit the size of ducts to pass through them while Vierendeel girders allow larger ducts.

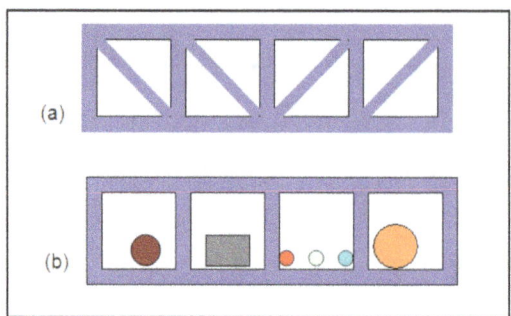

Truss (a) vs. Vierendeel (b) duct constraints

Weather and Climate Conditions

Weather conditions obviously will affect the selection of structural material. For example, in hot weather area like Saudi Arabia using of wood would not be recommended as much as using of concrete or steel , since wood is not available. In addition, the weather conditions include strong winds and snow. For instance, in high wind speed areas it is better to use heavyweight material or braced frame system or shear walls in order to prevent uplift and minimize the lateral drift caused by wind force. Hence, climate will affect the selection of structural system and material as well. In snow area it is better to use the sloped roofs to minimize snow load. To design for snow load will be necessary in order to prevent roof collapse.

Security

According to current world events most buildings are designed to protect occupants from external dangers. There are several methods to provide security for buildings. One of them is to select the safest structural material. Concrete is highly recommended in a valuable building that requires a high level of security like data processing centers, hospitals, military constructions, and nuclear facilities.

Exterior Cladding

Structural systems and materials should also be integrated with the exterior cladding. Several types of exterior cladding are available: precast concrete, cast in place concrete, masonry, glass, metal, stone, aluminum sheets, and marble. Cast in place concrete is one of the best but most costly systems because it can act as exterior cladding and structural frame as well. Furthermore, no need for connection between the structural element and the architectural concrete walls. The exterior cladding system should be known in the first structural stages in order to calculate their loads.

Minimal Story Height

Structural systems and materials should reduce story height as much as possible in order to reduce the total cost as well. However, high stories are preferred for luxury housing. Concrete structures require less story height than the other systems because flat plate floor slab does not require a structural beam at all which helps to pass the service lines under the slab. Reducing the exterior cladding height will minimize cost. In addition, reducing height will reduce costs for plumbing, electrical, and HAVC. Furthermore, decreasing floor height obviously will reduces building volume and then reduce HAVC consumption.

Cash Flow and Financing Costs

Structural systems and materials should be adapted to the owner demands and his ability to pay. For example, cost is different for cast in place and precast concrete. Steel fabrication requires pre-financing and long time of order, while concrete requires less time and less costs. Hence, concrete can save owner money, keeping money in the bank until the material is ready. In addition, concrete can be provided in shorter time which will help to finish the construction on time in comparison to steel which requires pre-order. For example, the Wall Street journal project costs $175 million for construction and the delay to finish this project cost $9 million per month ($3 million in interest and $6 million in lost rent) for about $300,000 a day.

Building Configuration

Since the structure system is defined by the form of the building, there is a strong correlation between structure system and building shape. The shape of building could be the shape of plan or elevation. A complex shape will cost more than a simple shape, and may require a specific material, labor skills, and technology.

Future Modification

Structural system should provide flexibility to allow future changes. Windows, doors, plan layout, etc. may change in the future and require the structure to allow such changes. Concrete floor slab like flat slab as well as flat plate slab are adoptive to this requirements.

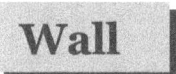

Wall construction is a Method for constructing walls for buildings. Walls are constructed in different forms and of various materials to serve several functions. Exterior walls protect the building interior from external environmental effects such as heat and cold, sunlight, ultraviolet radiation, rain and snow, and sound, while containing desirable interior environmental conditions. Walls are alsodesigned to provide resistance to passage of fire for some defined period of time, such as a onehour wall. Walls oftencontain doors and windows, which provide for controlled passage of environmental factors and people through the wall line.

Walls are designed to be strong enough to safely resist the horizontal and vertical forces imposed upon them, as defined bybuilding codes. Such loads include wind forces, selfweight, possibly the weights of walls and floors from above, the effects ofexpansion and contraction as generated by temperature and humidity variations as well as by certain impacts, and the wearand tear of interior occupancy.

Modern building walls may be designed to serve as either bearing walls or curtain walls or as a combination of both inresponse to the design requirements of the building as a whole. Both types may appear similar when complete, but theirsequence of construction is usually different.

Wall is a structure defining an exact area and providing safety & shelter. There are various types of walls used in the construction of buildings given below.

Types of Walls

- Load Bearing Walls
  - Precast Concrete Wall
  - Retaining Wall
  - Masonry Wall
  - Pre Panelized Load Bearing Metal Stud Walls
  - Engineering Brick Wall
  - Stone Wall
- Non-Load Bearing Wall
  - Hollow Concrete Block
  - Facade Bricks
  - Hollow Bricks
  - Brick Walls
- Cavity Walls
- Shear Walls
- Partition Walls
- Panel Walls
- Veneered Walls
- Faced Walls

Brief descriptions of different types of walls are given below:

Load Bearing Walls

Load bearing wall is a structural element. It carries the weight of a house from the roof and upper floors, all the way to the foundation. It supports structural members like beams(sturdy pieces of wood or metal), slab and walls on above floors above. A wall directly above the beam is called load bearing wall. It is designed to carry the vertical load. In another way, if a wall doesn't have any walls, posts or other supports directly above it, it is more likely to be a load-bearing wall. Load bearing walls also carry their own weight. This wall is typically over one another on each floor. Load bearing walls can be used as interior or exterior wall. This kind of wall will often be perpendicular to floor joists or ridge. Concrete is an ideal material to support these loads. The beams go directly into the concrete foundation. Load bearing walls inside the house tend to run the same direction as the ridge.

Types of Load Bearing Walls:

- Precast Concrete Wall

- Retaining Wall

- Masonry Wall

- Pre Panelized Load Bearing Metal Stud Walls

- Engineering Brick Wall

- Stone Wall

Non-Load Bearing Walls

A wall which doesn't help the structure to stand up and holds up only itself is known as a non-load bearing wall. It doesn't support floor roof loads above. It is a framed structure. Most of the time, They are interior walls whose purpose is to divide the structure into rooms. They are built lighter. One can remove any non-load bearing walls without endangering the safety of the building. Non-load bearing walls can be identified by the joists and rafters. They are not responsible for gravitational support for the property. It is cost effective. This wall is referred to as "curtain wall".

Types of Non-Load Bearing Wall:

- Hollow Concrete Block

- Facade Bricks

- Hollow Bricks
- Brick Walls

Cavity Walls

The cavity wall consists of two separate wythes. The wythes are made of masonry. Those two walls are known as internal leaf and external leaf. This wall is also known as a hollow wall. They reduce their weights on the foundation. They act as good as sound insulation. Cavity wall gives better thermal insulation than any other solid wall because space is full of air and reduces heat transmission. They have a heat flow rate that is 50 percent that of a solid wall. It is economically cheaper than other solid walls. It is fire resistant. Cavity wall helps to keep out from noise.

Shear Walls

It is a framed wall. It is designed to resist lateral forces. This lateral force comes from exterior walls, floor, and roofs to ground foundation. The usage of the shear wall is important, especially in large and high-rise buildings. It is Typically constructed from materials like concrete or masonry. It has an excellent structural system to resist earthquake. It provides stiffness in the direction. The construction and implementation are easy in shear walls. It is located symmetrically to reduce ill effects of a twist. Shear wall doesn't exhibit any stability problem.

Partition Walls

It is used in separating spaces from buildings. It can be solid, constructed from brick or stone. It is a framed construction. The partition wall is secured to the floor, ceiling, and walls. It is enough strong to carry its own load. It resists impact. It is stable and strong to support wall fixtures. Partition wall works like a sound barrier and it is fire resistant.

Panel Walls

It is a non-bearing wall between columns or pillars that are supported. The panel is installed with both nails and adhesive. The paneling design choices include rustic, boards, frame. Paneling can be from hardwoods or inexpensive pine. One should paint the space before installing panel walls.

Veneered Walls

With a veneered wall, we are holding up the material. It can be made of brick or stone. The most famous veneered wall is made of brick. The wall is only one wythe thick. It became the norm when building codes began to require insulation in the interior walls. It is light weighted. The construction takes less time to complete in veneered walls.

Faced Walls

It is a wall which masonry facing and backing are so bonded as to exert common action under load. It creates a streamlined look. The faced wall is easy to install.

Flooring

Flooring is the process of providing clean, smooth, durable and impervious levelled surface to the occupants of the house and the said surface is called Floor. The Floor should be hard, durable and sufficiently strong to withstand the loads coming over it.

Process of Flooring:

All the floors consist of two main components

i.   Floor base: At ground level the floor base consists of layers of various materials. The purpose of the floor base is to provide a strong and unsinkable surface to the floor covering. First of all the filled up soil is watered and rammed properly. Above it a layer of boulders or broken bricks of at least 300mm is provided and rammed properly. Now a layer of 150mm thick cement concrete is laid to provide a level surface for the floor covering. Other than ground floor i.e. 1st floor, 2nd floors etc. the roof slab itself acts as a floor base.

ii. Floor covering: This is a layer of materials like Tiles, Mosaic, Marble etc. laid over the floor base. The name of the floor is determined by the materials used for floor covering. The choice of flooring material depends upon several factors. Commonly Cement concrete flooring is used most of the houses. But now a days Tiled Flooring, Marble Flooring and PVC flooring are becoming a hot trend in modern houses as they provide trendy look to the flooring.

Types of Flooring

Hardwood Flooring

Hardwood is coveted for its natural, timeless beauty and lasting durability. Choose from two hardwood flooring types: solid hardwood planks are 100% solid hardwood, and engineered wood has a stable, layered construction with a wood top and backing.

Places where Hardwood can be Installed

Hardwood flooring can be installed in most rooms of your home, though we do not recommend it for bathrooms, laundry rooms or any other areas that could be subject to puddles or high humidity. However, engineered wood can be a good choice for basement installations or over concrete slab or radiant heating systems.

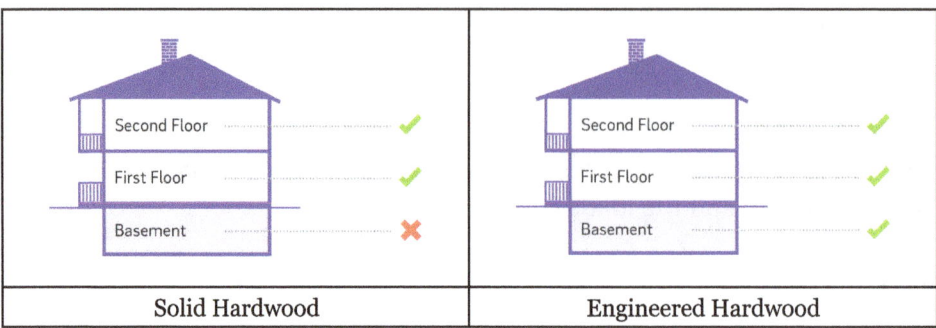

| Solid Hardwood | Engineered Hardwood |

Durability of Hardwood

Even highly durable hardwood can show wear over time, but generally this flooring type offers the long-lasting performance that floor shoppers value. The hardness of the wood species, level of protection of the finish, and how much traffic your floor gets will all affect how well a particular hardwood floor holds up against scratches, dents and stains.

Kinds of Styles with Hardwood

Armstrong Flooring offers an incredible selection of hardwood flooring styles, from elegant to rustic. Our collections feature a full range of domestic, imported and exotic wood species, colors, finishing effects, textures and plank widths. Whatever your style, you're sure to find a look you'll love.

Luxury Vinyl Flooring

Also known as luxury vinyl tile (LVT), this type of flooring comes in wood and stone looks that are designed to resist moisture and everyday wear and tear in the most active homes.

Places where Luxury Vinyl Can be Installed

LVT is the perfect type of flooring for bathrooms, kitchens, basements and laundry rooms, but it can be installed anywhere in your home. There are even options that allow you to easily install over subfloors with minor irregularities.

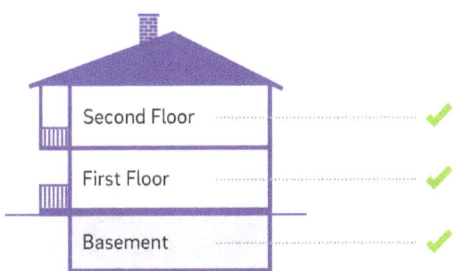

Durability of Luxury Vinyl

Luxury vinyl is very durable and easy to clean. In fact, our luxury vinyl collections are 100% waterproof, so they won't swell, buckle or lose integrity when exposed to water. And while all LVT products are scratch, stain and wear resistant, you can choose between different performance levels. Some come with the superior scratch-and-stain resistance of Diamond 10 Technology, making them tough enough to be backed by our lifetime warranty.

Styles Available with Luxury Vinyl Flooring

The kind of trends you might see in hardwood planks or natural stone tiles can often be found in luxury vinyl. Traditional, rustic and exotic woods in plank sizes as wide as 7". Travertine or marble-mimicking tiles. Modern industrial designs inspired by concrete or petrified wood. So many options means it's easy to find a floor that does what you need it to do, but is really stylish, too.

Engineered Tile Flooring

This type of flooring is engineered with 70% limestone and other resilient materials to be the perfect alternative to natural stone, porcelain or ceramic tile. It has the character and longevity you love about traditional tile, but it's warmer and softer under-foot.

Places where Engineered Tile can be Installed

Engineered tiles can be installed anywhere in the home, with or without grout. It's ideal for high-moisture areas like kitchens, bathrooms, basements and laundry rooms — and even on the wall as a backsplash or accent wall! And unlike ceramic tile, engineered tile can be installed over minor subfloor irregularities without the risk of cracking.

Durability of Engineered Tile

At Armstrong Flooring, our engineered tile styles are all backed by our lifetime warranty. This flooring type is built to handle all the things you need your tile floors to handle: high traffic, spills, messes and dirt. Even the grout is stain resistant. And if you drop heavy objects on engineered tile, it's not vulnerable to cracks like natural stone can be.

Styles Available with Engineered Tile

There are many styles you can achieve with engineered tile, starting with classic slate, terracotta, travertine and marble — all the way to tiles that resemble reclaimed wood and linen fabric. Explore pattern possibilities, on-the-wall looks and how different tile sizes and shapes can alter the look of your space in stunning ways.

Rigid Core Flooring

Maximum durability and award-winning design are the hallmarks of rigid core flooring. It has a layered structure that features the best attributes of multiple flooring types, including luxury vinyl and laminate. Rigid core flooring features many of the durability characteristics of traditional luxury vinyl tile, but with greater dent resistance and design realism.

Places where Rigid Core Flooring can be Installed

Enjoy this flooring type in any room of your home. It's durable enough for the highest traffic and most moisture-prone rooms, and beautiful enough with its realistic hardwood designs to elevate the style of your main living spaces and bedrooms.

Durability of Rigid Core Flooring

Rigid core flooring, unlike many other types of flooring, offers the ultimate dent resistance against things like dropped objects and high heels. It's also 100% waterproof, so when exposed to water the planks will not swell, buckle or lose integrity.

Kinds of Styles available with Rigid Core Flooring

Our award-winning designs include highly realistic visuals and textures that resemble real natural hardwood. Capture the look of traditional wood floors or trending rustic, reclaimed and brushed wood looks.

Vinyl Sheet Flooring

Duality Premium Collection - Eclipse

Vinyl sheet is a resilient type of flooring that comes in a large roll, cut to size.

Places Where Vinyl Sheet Flooring can be Installed

Vinyl sheet is durable enough to be installed anywhere in the home, and its single solid surface makes it a good choice for moisture-prone kitchens and baths.

Durability of Vinyl Sheet Flooring

We offer three performance levels for vinyl sheet, with different durability and comfort features. Our two highest performing vinyl sheet collections, Duality Premium and CushionStep Better, feature Diamond 10® Technology for superior scratch, scuff and stain resistance.

Different styles of Vinyl Sheet Flooring

Vinyl sheet is often turned to as an alternative flooring type for homeowners who want the look of natural stone, ceramic tile or even hardwood, for less. Cutting-edge print technology produces realistic limestone, sandstone, slate, traditional hardwood and exotic hardwood looks, as well as alternative styles and patterned designs like black-and-white checkerboard.

Vinyl Tile

Vinyl tile is a great type of flooring for do-it-yourself projects. With peel-and-stick installation, it's easy to transform your room over a weekend. Like vinyl sheet, tile is durable, affordable and low-maintenance.

Terraza Grand - Arctic White

Places where Vinyl Tile Flooring can be Installed

This flooring type can be installed on any flat and level surface besides stairs, and is typically used for kitchens, bathrooms, mudrooms and playrooms.

Durability of Vinyl Tile Flooring

Under normal household use, vinyl tile holds up well against foot traffic, kids and dropped objects. Vinyl tile flooring is warranted for 5 to 10 years.

Different styles of Vinyl Tile Flooring

Most vinyl tile comes in 12" x 12" square tiles and you can select from looks that resemble stone tile, porcelain tile and parquet wood, as well as classic geometric and checkerboard patterns.

Laminate

With this floor type you can get the look and feel of exotic wood or high-end stone on a practical budget. DIYers will find Lock&Fold laminate a breeze to install.

Places where Laminate Flooring can be Installed

You can install this type of flooring almost anywhere: bathrooms, kitchens, mudrooms, foyers, dining rooms, living rooms, family rooms and bedrooms.

Durability of Laminate Flooring

Laminate is designed to withstand everyday wear from the most active households, and — depending on the product you choose — even commercial spaces like main street businesses. It has a layered construction that includes a hardened wear layer to protect from scratches, surface spills, stains and fading, and an inner core that adds moisture resistance and stability.

Different Styles of Laminate Flooring

Whatever your design vision, it will be easy to achieve with laminate, and you can be sure it will look authentic, too. Hardwood styles include rustic, weathered, hand-scraped and reclaimed looks. Stone looks range from Spanish-style pavers to slate tile. And our innovative design technology even allows us to create unique flooring inspired by unexpected materials, blending wood, metal and concrete visuals.

Ventilation

Ventilation moves outdoor air into a building or a room, and distributes the air within the building or room. The general purpose of ventilation in buildings is to provide healthy air for breathing by both diluting the pollutants originating in the building and removing the pollutants from it.

Building ventilation has three basic elements:

- *ventilation rate* — the amount of outdoor air that is provided into the space, and the quality of the outdoor air;

- *airflow direction* — the overall airflow direction in a building, which should be from clean zones to dirty zones; and

- *air distribution or airflow pattern* — the external air should be delivered to each part of the space in an efficient manner and the airborne pollutants generated in each part of the space should also be removed in an efficient manner.

Assessing Ventilation Performance

Ventilation performance in buildings can be evaluated from the following four aspects, corresponding to the three basic elements of ventilation discussed above.

- Does the system provide sufficient ventilation rate as required?

- Is the overall airflow direction in a building from clean to dirty zones (e.g. isolation rooms or areas of containment, such as a laboratory)?

- How efficient is the system in delivering the outdoor air to each location in the room?

- How efficient is the system in removing the airborne pollutants from each location in the room?

Two overall performance indices are often used. The air exchange efficiency indicates how efficiently the fresh air is being distributed in the room, while the ventilation effectiveness indicates how efficiently the airborne pollutant is being removed from the room. Engineers define the local mean age of air as the average time that the air takes to arrive at the point it first enters the room, and the room mean age of air as the average of the age of air at all points in the room. The age of air can be measured using tracer gas techniques.

The air exchange efficiency can be calculated from the air change per hour and the room mean age of air. For piston-type ventilation, the air exchange efficiency is 100%, while for fully mixing ventilation the air exchange efficiency is 50%. The air exchange efficiency for displacement ventilation is somewhere in between, but for short-circuiting the air exchange efficiency is less than 50%.

Ventilation effectiveness can be evaluated by either measurement or simulation. In simple terms, the ventilation flow rate can be measured by measuring how quickly injected tracer gas is decayed in a room, or by measuring the air velocity through ventilation openings or air ducts, as well as the flow area. The airflow direction may be visualized by smoke. Computational fluid dynamics and particle image velocimetry techniques allow the air distribution performance in a room to be modelled.

Ventilation Types

Natural Ventilation

Natural ventilation is the use of wind and temperature differences to create airflows in and through buildings.

There are two basic types of natural ventilation effects: buoyancy and wind. Buoyancy ventilation is more commonly referred to as temperature-induced or stack ventilation. Wind ventilation supplies air from a positive pressure through openings on the windward side of a building and exhausts air to a negative pressure on the leeward side. Airflow rate depends on the wind speed and direction as well as the size of openings. In summer, the indoor-outdoor temperature difference is not high enough to drive buoyancy ventilation, and wind is used to supply as much fresh air as possible. In winter, however, the indoor is much warmer than outdoors, providing an opportunity for buoyancy ventilation.

Task Ventilation

Traditional ventilation systems supply a mixture of outside and re-circulated air in high velocity jets so that the indoor air in rooms is often well mixed. This can be an inefficient method of delivering outside air to an occupant. Task-ambient conditioning (TAC) systems are a ventilation technology with the potential for improved ventilation to the occupant. TAC systems may supply air from the floor, desk, or partitions and enable occupants to adjust the supply flow rate, direction, or temperature so that thermal conditions can be tailored to meet the individual's requirements.

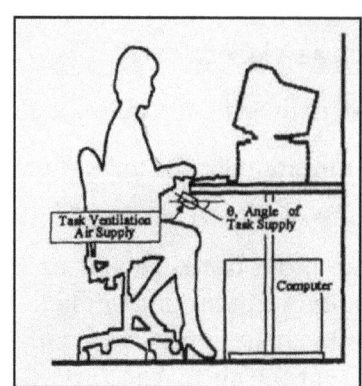

Task or personalized ventilation is a method for providing occupants with control of a local supply of air so that they can adjust their individual thermal environment. Controlled variables could be the supply-air temperature, velocity, direction, the ratio of room air to outside in the supply air, and the radiant temperature. These systems may provide all or part of the conditioned air to the occupied space. Task or personalized ventilation systems also have the potential to improve ventilation at the occupant's breathing zone because they can provide supply air preferentially toward the breathing zone. Supply air from these systems usually contains a high percentage of outside air, which generally does not contain a high concentration of indoor-generated pollutants. The air supply outlets of current task or personalized ventilation systems are located at the floor, mounted on the desk, or incorporated within the workstation partitions.

Mechanical Ventilation

This system supplies the required air flow at a constant rate. Ventilation is supplied by forcing air through the ducting with the use of a fan. The use of the fan however uses a lot of energy and consequently greater CO_2 emissions.

Hybrid Ventilation

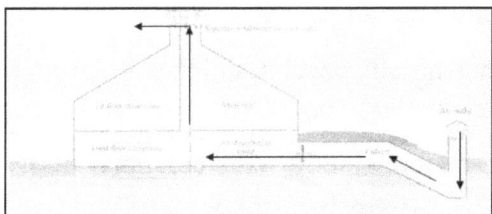

Hybrid ventilation is the mix of natural and mechanical ventilation. In this project there is only one aspect of mechanical ventilation, which contributes to the hybrid one: the fan which enhances the natural stack effect if the conditions are poor.

Comparison of Mechanical and Natural Ventilation

Mechanical Ventilation

If well designed, installed and maintained, there are a number of advantages to a mechanical system:

- Mechanical ventilation systems are considered to be reliable in delivering the designed flow rate, regardless of the impacts of variable wind and ambient temperature. As mechanical ventilation can be integrated easily into air-conditioning, the indoor air temperature and humidity can also be controlled.

- Filtration systems can be installed in mechanical ventilation so that harmful microorganisms, particulates, gases, odours and vapours can be removed.

- The airflow path in mechanical ventilation systems can be controlled, for instance allowing the air to flow from areas where there is a source (e.g. patient with an airborne infection), towards the areas free of susceptible individuals.

- Mechanical ventilation can work everywhere when electricity is available.

However, mechanical ventilation systems also have problems.

- Mechanical ventilation systems often do not work as expected, and normal operation may be interrupted for numerous reasons, including equipment failure, utility service interruption, poor design, poor maintenance or incorrect management. If the system services a critical facility, and there is a need for continuous operation, all the equipment may have to be backed up — which can be expensive and unsustainable.

- Installation and particularly maintenance costs for the operation of a mechanical ventilation system may be very high. If a mechanical system cannot be properly installed or maintained due to shortage of funds, its performance will be compromised.

Because of these problems, mechanical ventilation systems may result in the spread of infectious diseases through health-care facilities, instead of being an important tool for infection control.

Natural Ventilation

If well installed and maintained, there are several advantages of a natural ventilation system, compared with mechanical ventilation systems.

- Natural ventilation can generally provide a high ventilation rate more economically, due to the use of natural forces and large openings.

- Natural ventilation can be more energy efficient, particularly if heating is not required.

- Well-designed natural ventilation could be used to access higher levels of daylight.

From a technology point of view, natural ventilation may be classified into simple natural ventilation systems and high-tech natural ventilation systems. The latter are computer-controlled, and may be assisted by mechanical ventilation systems (i.e. hybrid or mixed-mode systems). High-tech natural ventilation may have the same limitations as mechanical ventilation systems; however, it also has the benefits of both mechanical and natural ventilation systems.

If properly designed, natural ventilation can be reliable, particularly when combined with a mechanical system using the hybrid (mixed-mode) ventilation principle, although some of these modern natural ventilation systems may be more expensive to construct and design than mechanical systems.

In general, the advantage of natural ventilation is its ability to provide a very high air-change rate at low cost, with a very simple system. Although the air-change rate can vary significantly, buildings with modern natural ventilation systems (that are designed and operated properly) can achieve very high air-change rates by natural forces, which can greatly exceed minimum ventilation requirements.

There are a number of drawbacks to a natural ventilation system.

- Natural ventilation is variable and depends on outside climatic conditions relative to the indoor environment. The two driving forces that generate the airflow rate (i.e. wind and temperature difference) vary stochastically. Natural ventilation may be difficult to control, with airflow being uncomfortably high in some locations and stagnant in others. There is a possibility of having a low air-change rate during certain unfavourable climate conditions.

- There can be difficulty in controlling the airflow direction due to the absence of a well-sustained negative pressure; contamination of corridors and adjacent rooms is therefore a risk.

- Natural ventilation precludes the use of particulate filters. Climate, security and cultural criteria may dictate that windows and vents remain closed; in these circumstances, ventilation rates may be much lower.

- Natural ventilation only works when natural forces are available; when a high ventilation rate is required, the requirement for the availability of natural forces is also correspondingly high.

- Natural ventilation systems often do not work as expected, and normal operation may be interrupted for numerous reasons, including windows or doors not open, equipment failure (if it is a high-tech system), utility service interruption (if it is a high-tech system), poor design, poor maintenance or incorrect management.

- Although the maintenance cost of simple natural ventilation systems can be very low, if a natural ventilation system cannot be installed properly or maintained due to a shortage of funds, its performance can be compromised, causing an increase in the risk of the transmission of airborne pathogens.

These difficulties can be overcome, for example, by using a better design or hybrid (mixed-mode) ventilation. Other possible drawbacks, such as noise, air pollution, insect vectors and security, also need to be considered. Because of these problems, natural ventilation systems may result in the spread of infectious diseases through health-care facilities, instead of being an important tool for infection control.

Types of Ventilation Systems

Exhaust Ventilation Systems

Exhaust ventilation systems work by depressurizing the building. By reducing the inside air pressure below the outdoor air pressure, they extract indoor air from a house while make-up air infiltrates through leaks in the building shell and through intentional, passive vents.

Exhaust ventilation systems are most applicable in cold climates. In climates with warm, humid summers, depressurization can draw moist air into building wall cavities, where it may condense and cause moisture damage.

Exhaust ventilation systems are relatively simple and inexpensive to install. Typically, an exhaust ventilation system is composed of a single fan connected to a centrally located, single exhaust point in the house.

A preferable option is to connect the fan to ducts from several rooms (especially rooms where pollutants tend to be generated, such as bathrooms). Adjustable, passive vents through windows or walls can be installed to introduce fresh air rather than rely on leaks in the building envelope. However, passive vents may be ineffective because larger pressure differences than those induced by the ventilation fan may be needed for them to work properly.

Spot ventilation exhaust fans installed in the bathroom but operated continuously represent an exhaust ventilation system in its simplest form.

One concern with exhaust ventilation systems is that they may draw pollutants, along with fresh air, into the house. For example, in addition to drawing in fresh outdoor air, they may draw in the following:

- Radon and molds from a crawlspace

- Dust from an attic

- Fumes from an attached garage

- Flue gases from a fireplace or fossil fuel–fired water heater and furnace.

This can especially be of concern when bath fans, range fans, and clothes dryers (which also depressurize the home while they operate) are run when an exhaust ventilation system is also operating.

Exhaust ventilation systems can also contribute to higher heating and cooling costs compared with energy recovery ventilation systems because exhaust systems do not temper or remove moisture from the make-up air before it enters the house.

Supply Ventilation Systems

Supply ventilation systems work by pressurizing the building. They use a fan to force outside air into the building while air leaks out of the building through holes in the shell, bath- and range-fan ducts, and intentional vents.

As with exhaust ventilation systems, supply ventilation systems are relatively simple and inexpensive to install. A typical system has a fan and duct system that introduces fresh air into usually one—but preferably several—rooms that residents occupy most (for example, bedrooms, living room, kitchen). This system may include adjustable window or wall vents in other rooms.

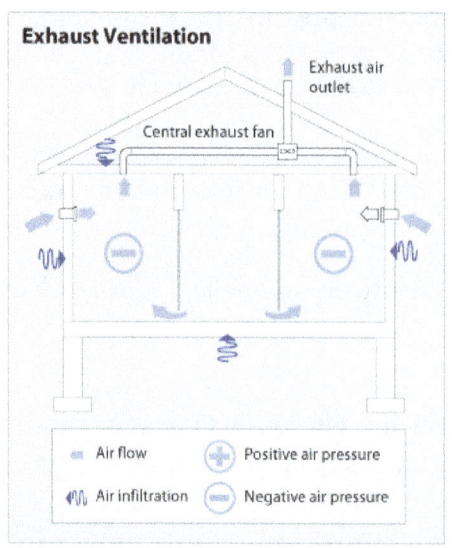

Supply Ventilation System

Supply ventilation systems allow better control of the air that enters the house than do exhaust ventilation systems. By pressurizing the house, these systems discourage the entry of pollutants from outside and prevent backdrafting of combustion gases from fireplaces and appliances. They also allow air introduced into the house to be filtered to remove pollen and dust or to be dehumidified.

Supply ventilation systems work best in hot or mixed climates. Because they pressurize the house, they have the potential to cause moisture problems in cold climates.

In winter, the supply ventilation system causes warm interior air to leak through random openings in the exterior wall and ceiling. If the interior air is humid enough, some moisture may condense in the attic or parts of the exterior wall, where it can promote mold, mildew, and decay.

Like exhaust ventilation systems, supply ventilation systems do not temper or remove moisture from the air before it enters the house. Thus, they may contribute to higher heating and cooling costs compared with energy recovery ventilation systems. Because air is introduced in the house at discrete locations, outdoor air may need to be mixed with indoor air before delivery to avoid cold air drafts in winter. An in-line duct heater is another option, but it will increase operating costs.

Balanced Ventilation Systems

Balanced ventilation systems, if properly designed and installed, neither pressurize nor depressurize a house. Rather, they introduce and exhaust approximately equal quantities of fresh outside air and polluted inside air, respectively. A balanced ventilation system usually has two fans and two duct systems. It facilitates good distribution of fresh air by placing supply and exhaust vents in appropriate places.

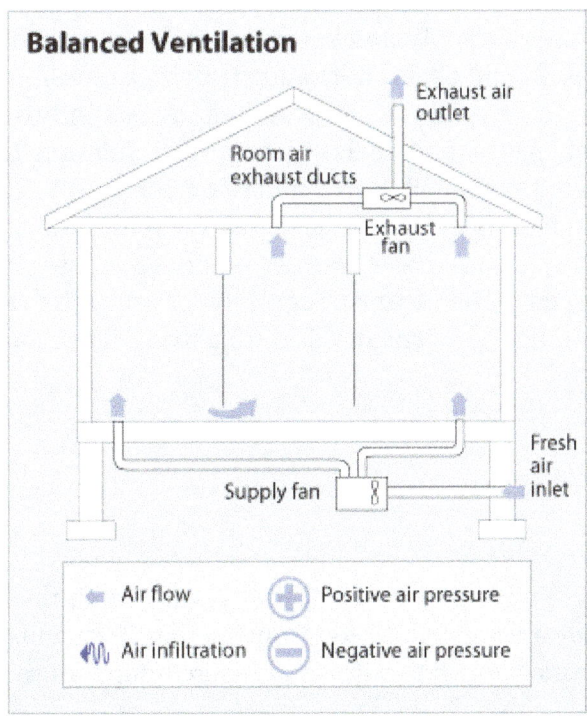

A typical balanced ventilation system is designed to supply fresh air to bedrooms and common rooms where people spend the most time. It also exhausts air from rooms where moisture and pollutants are most often generated, such as the kitchen, bathrooms, and the laundry room.

Like both supply and exhaust systems, balanced ventilation systems do not temper or remove moisture from the air before it enters the house.

They do, however, use filters to remove dust and pollen from outside air before introducing it into the house.

Also, like supply ventilation systems, outdoor air may need to be mixed with indoor air before delivery to avoid cold air drafts in the winter. This may contribute to higher heating and cooling costs.

Balanced ventilation systems are appropriate for all climates; however, because they require two duct and fan systems, they are usually more expensive to install and operate than supply or exhaust systems.

Energy Recovery Systems

Energy recovery ventilation systems usually cost more to install than other ventilation systems. In general, simplicity is key to a cost-effective installation. To save on installation costs, many systems share existing ductwork.

Complex systems are not only more expensive to install, but often they are also more maintenance intensive and consume more electric power. For most houses, attempting to recover all of the energy in the exhaust air will probably not be worth the additional cost. Also, these types of ventilation systems are still not very common. Only some HVAC contractors have enough technical expertise and experience to install them.

In general, you want to have a supply and return duct for each bedroom and for each common living area. Duct runs should be as short and straight as possible. The correct size duct is necessary to minimize pressure drops in the system and thus improve performance. Insulate ducts located in unheated spaces, and seal all joints with duct mastic. Also, energy recovery ventilation systems operated in cold climates must have devices to help prevent freezing and frost formation. Very cold supply air can cause frost formation in the heat exchanger, which can damage it. Frost buildup also reduces ventilation effectiveness. In addition, energy recovery ventilation systems need to be cleaned regularly to prevent deterioration of ventilation rates and heat recovery, and to prevent mold and bacteria from forming on heat exchanger surfaces.

Roofing

Roofs take quite a beating. Fully faced toward the sky, they catch the brunt of weather's worst. They have to be able to take a licking and keep from leaking. They must be weathertight, secure, durable, attractive, and elastic enough to withstand severe temperature shifts without cracking.

Carpenter uses a pneumatic nailer to attach plywood sheathing to the roof
rafters—in this case, a series of roof trusses.

Over the centuries, roof-building techniques have been refined to yield roofs of considerable
strength and durability. A wide variety of materials has been developed that will last many years—
in some cases, as long as the house. And homeowners have a vast selection of materials, colors,
prices, and other features from which to choose.

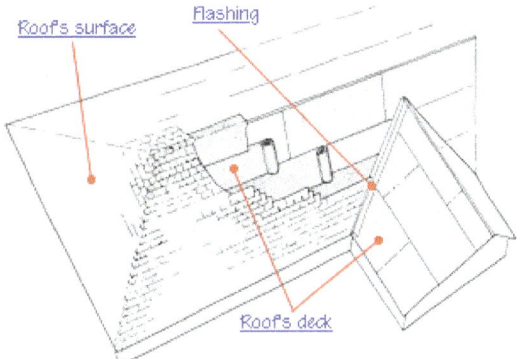

Roof Construction Diagram

A contemporary roof, regardless of shape or surface material, consists of a variety of components that
include wood framing, sheathing, underlayment, flashing, gutters, and, of course, the shingles or other
finished surface.

The basics of Building a Roof System

A common misunderstanding in roofing is that the shingles is the most important part; when in
reality each part of the system is equally important. Since you cannot see all of the components,
this is where many contractors "cut corners" to offer cheap prices. At Tuttle Contracting we not
only strive to provide superior service, but also to install a premium roof; never cutting corners,
and using premium grade products.

1. After the roof is removed, new underlayment must be installed over the sheathing. The underlayment should be attached to the roof using "plastic cap" nails and each course should
   overlap the next by three to four inches. As with shingles there is a variety of types of underlayment to choose from. For years tar saturated felt-paper, commonly called "felt" or "tar paper" has been the standard in roofing underlayment and is still the most used product today.

2. In cold weather climates an ice barrier is needed in addition to the underlayment. An ice barrier does what standard underlayment doesn't because it has an adhesive backing that adheres to the sheathing and creates a water tight seal. Local code requirements will dictate the necessity of having an ice barrier. But, simply stated it is only needed if snow and ice commonly accumulate on the roof thereby creating the possibility of "ice damming." Ice damming occurs when ice builds up along the edge of the roof making it possible for water to back-flow under the shingles and into the house.

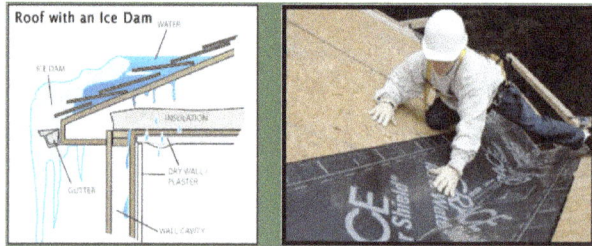

3. Once the proper underlayment has been installed, all vents and pipe jacks should be installed (with the exception of ridge vent which is put on after the shingles). While there are a variety of options, all of the vents and pipe jacks are specifically designed to cover and protect penetrations through the roof. But that doesn't make all vents equal.

4. At this point, you could say the roofing surface has been prepared and you are now ready to begin installing the shingles. The shingles will be installed from the bottom up and

will follow a repeating diagonal pattern with a staggered joint pattern. Each shingle must be properly nailed along the designated "nail line" with the nails penetrating at least ¼" through the decking. When done properly with an attention to detail it will result in a waterproof roof that is both functional and cosmetically appealing.

5. Once the body of the roof is fully covered with shingles, then the ridge caps must be installed. If the roof design calls for ridge vent, then the ridge vent will also be installed at this time with the ridge shingles being nailed over them. Care should be taken to ensure the ridge lines are straight and that any exposed nail heads are covered with a UV resistant sealant to ensure a water-tight surface.

6. At Tuttle Contracting the roof is not considered finished until all of the little details are complete. All fixtures on the roof will be painted to blend and match with the color of the roof you have chosen, the roof will be blown off, your gutters will be cleaned out, and the roof will receive a final inspection by both the Crew Leader.

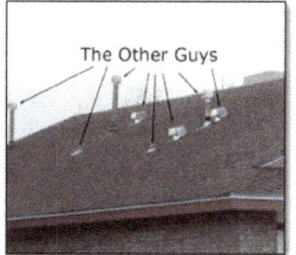

7. On the ground, everything will be picked up and transported off of your property. Care will be taken to protect and clean around sensitive areas, such as flower beds. Hard surfaces will be swept or blown off, the lawn will be raked to pick-up the little pieces as well as the large ones, and large magnets will be used to ensure nails are not left around your home.

The roofing system is a much more complex set of materials and engineering than most people give it credit for. Different components all work together to keep your home safe from the weather, elements and even some man made harm. One of the most important aspects of the roof is the roof covering, so what is roof covering? Let's look at what the roof covering is designed to do and some of the different material choices.

Roof Covering

Roof covering is what most people think of when someone says the roof. While the roofing system is composed of several components, the roof covering is the most visual part of the whole system and makes up the majority of waterproofing and protection. Selection of a proper roof covering is one of the most important aspects of getting your roof redone.

Types of Roofing Material

Roofs have been around ever since humans started living in shelters so you can imagine there are several different types of roof covering available all around the world. The most popular roof covering in the entire world is roofing tiles while here in the United States, asphalt shingles dominate the market. These are just two of the most popular roof coverings. Other popular materials include slate tiles, metal roofing, wood shakes and even green roofs in certain parts of the world.

Roof Decking

The roof deck is the roofing material between the structural components (the trusses and joists) and the insulation and weatherproofing layers (roof materials, coatings, layers, etc.). In short, the roof deck is the section of the roof onto which everything else is placed. As such it needs to be strong enough to hold the weight and durable enough to cope with some give.

Types of deck material include steel, concrete, cement and wood. The type of roof deck to use depends on how much weight it needs to carry, which further depends on the roofing materials selected. Other factors include the weight of air conditioning equipment, rain and snow. Special features such as walking decks, roof top swimming pools and bars will also need extra support.

For residential buildings, the most common type of roof deck is plywood or tongue and groove wood systems. Wood generally stands up to the weight of almost any type of roofing material placed on top, although tile may require extra reinforcement.

For commercial or industrial buildings, steel, concrete or cement are the more frequently chosen roof deck materials. That is because the large structures tend to span more distance and carry more weight in equipment on the roof, including air conditioning, air circulation pumps, etc. Since the equipment must be serviced frequently, the roof deck has to be able to take the weight of maintenance personnel on the roof on a regular basis, as well as everything the elements can throw at it.

In all cases, a professional roofing expert should be used to install the roof deck. Since the deck must hold the weight of everything else, improper installation can be dangerous. In addition, a professional roofer has the expertise to select the proper deck material to hold the load of your roofing materials and the various systems and components that operate from the roof.

Ceiling

The ceiling of a building form one of the most important structural elements it terms of functionality as well as creativity which in turn bring good aesthetics to the building interior. This document would help you understand ceilings that in turn would help you understand the most suitable options while undertaking either repair or renovation work of our building interior.

The ceiling gains a definition," A part of a building which encloses and is exposed overhead in a room, protected shaft or circulation space."

It helps us to create an enclosure and separation between spaces. They provide perfect lighting in the room by controlling diffusion of light. It also controls the sound around a room by making the room sound proof. It, therefore, stops the passage of sound between the rooms.

Now ceiling also possesses fire resistant properties that would facilitate accommodation of building services such as vents, lighting, sprinkler heads etc.

Types of Ceiling used in Building Construction

1. Exposed Ceilings

As shown in below figure, this kind of ceiling arrangement would completely expose the structural and mechanical components of the building thus omitting a concept of finished ceiling. This arrangement lacks a discomfort in aesthetics but gains many advantages like the economy, easy maintenance due to ease of access. This also enables the thermal mass of the building to be exposed.

Further, the above-mentioned exposed thermal mass can be exploited by installing heating or cooling elements like chilled beams that will be discussed in the following discussions.

Figure. Exposed ceiling example in the interior of an office room

The disadvantage mentioned before of looks can vanish if we design , install roof structures and floor structures properly , leaving it exposed in the space below by means of timber beams or concrete slabs or space trusses. It's true that the mechanical element arrangement at ceiling level in a well systematic manner would create an attractive aesthetic effect.

2. Tightly Attached Ceiling

Ceilings of any material may be attached tightly to wood joists, wood rafters, steel joists or concrete slabs.

Special finishing arrangements must be worked out for any beams and girders that protrude through the plain of the ceiling and for ducts, conduits, pipes and sprinkler heads that fall below the ceiling.

3. Interstitial Ceilings

An interstitial space is defined as an intermediate space kept between regular-use floors. It is commonly located in hospitals and laboratory-type buildings to allow space for the mechanical systems of the building.

Laboratory and hospital rooms are easily rearranged throughout their lifecycles and therefore reduce lifecycle cost. This ensures more flexibility in the interiors of the building rooms.

They include a walkway for access with a low height. Mainly employed in buildings that have the following units:

- Electrical and communications wiring

- Air-conditioning ducts

- Oxygen

- Chilled water

- Vacuum pipe work

- Chemical sewer pipe work

- Water and waste pipework

- Fuel gas lines

- Compressed air lines

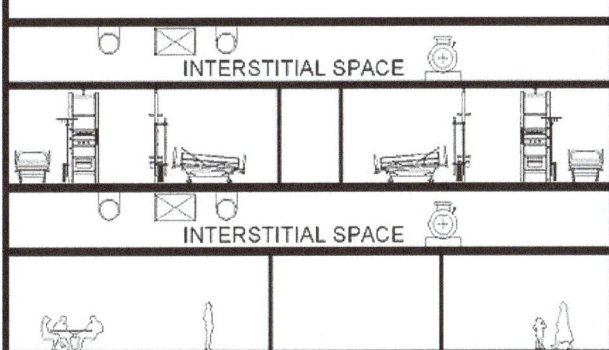

Figure: A picture depicting interstitial space arrangement

Figure: A Real Example of Interstitial Ceiling Space Arrangement

The ducts and pipe work occupy some space, which often requires continual maintenance and are subject to frequent change. The main advantage of interstitial ceilings is that they allow for maintenance and updating work, without interruption of activities in the spaces above and below.

4. Acoustical Ceilings

This ceiling is mainly implemented to control sound reverberation in a hall where there is the possibility of large sound propagation to make the hearing comfort zone. The acoustic ceiling material is made from fibrous materials that would absorb sound energy unlike other materials like plaster or gypsum ceilings.

These materials do not reduce transmission of sound between the spaces. They result in a reduction in the amount of sound that reflects into space. Hence it can be used to bring the modify acoustic character of space.

Figure. A Real Example of Acoustical Ceiling Arrangement

The performance of the ceiling material in terms of sound absorption can be expressed in terms of noise reduction coefficient (NRC). An NRC of 0.85 means that a ceiling material absorbs 85% o the sound that reaches it, and an almost 15% reflects back to the room.

Most of the acoustical ceilings NRC range from 0.5 to 0.9. This value is found to be 0.10 for plaster and gypsum ceiling board materials.

5. Radiant Chilled Ceilings

Radiant chilled ceilings include a network of chilled water coils in ceiling panels with insulation above. For some systems, pipework may be incorporated into plaster board. But this is found less efficient as plaster is an insulator.

The ceiling surface then cools the occupied space by both radiation and convection, providing temperatures throughout the space and avoids draught. The space requirement for chilled ceilings is found to be less, which may be installed with a depth of just 100mm. For some, a small-bore cooling coils can be embedded in plaster ceiling.

6. Convective Chilled Ceilings

These types of ceilings are a deviation from radiant chilled ceilings, in which the network of chilled water pipes incorporates fins, increasing the proportion of cooling that is provided by convection.

7. Suspended Ceilings

These are secondary ceilings suspended from the structure above (typically a floor or roof slab), creating a void between the underside of the slab and the top of the suspended ceiling. This void can provide a useful space for the sprinklers, distribution of heating, ventilation, and air conditioning (HVAC) services and plumbing and wiring services.

This also provides a display place for the installation of speakers, and smoke detectors, motion detectors, light fittings, wireless, antenna, CCTV, fire and so on. It provides an air 'plenum', in which the void itself forms a pressurized 'duct' to supply air to or extract it from the taken space below.

Figure: Suspended Ceilings

This is the one that contributes to fire-resistance in commercial and residential construction. In the case of a dropped ceiling (the other name of suspended ceiling), the rating is achieved by the entire system, which includes:

- The structure above, from which the ceilings is suspended. This can be a concrete floor or a timber floor

- The suspension mechanism

- The lowest membrane or dropped ceiling.

Between the structure that the dropped ceiling is suspended from and the dropped membrane, there is frequently some room for mechanical and electrical piping, wiring and ducting to be incorporated.

An independent ceiling can be constructed which has a separate fire-resistance. Such systems must be checked without the benefit of being suspended from a slab above so that it can hold itself up. This type of ceiling would be installed to protect items above from fire.

Plumbing System

Plumbing System is a system of pipes and fixtures installed in a building.

The plumbing in any building serves two main purposes. The first is to bring water into the structure for human use, and the second is to remove wastewater of various types. There are three main types of plumbing systems: potable water, sanitary drainage and stormwater drainage.

Potable Water System

The potable water system brings water into a structure; this water comes from the community water main. There is a valve on the water main itself for each structure that can be used to shut off its water supply. From there, a single pipe brings water into the structure, and it is then distributed to individual fixtures through a network of pipes. A meter keeps track of how much total water enters the structure.

Sanitary Drainage System

The sanitary drainage system removes wastewater from a building. It consists of pipes that take out human waste and fecal matter as well as wastewater from cooking, laundry, etc. The sanitary drainage system is connected to a series of vent pipes that go through the roof vertically; this allows for the venting of gases and for the entire system to operate at atmospheric pressure. The sanitary drainage system ultimately takes wastewater to the community sewer system.

Stormwater Drainage System

The purpose of the stormwater drainage system is to carry rainwater away from a structure. In older structures, rainwater simply drains into the sanitary drainage system, but in buildings that are more modern a separate system of drains carries water into the community storm sewers. Gutters are a part of the stormwater drainage system visible from outside the structure; other components, such as drains and pipes, are below the ground.

Types of Plumbing Systems in Buildings

Plumbing system in buildings consists of underground tank which is supplied water via municipal or water department supply lines, from there with the help of pumps and piping distribution system water is supplied to overhead tank and thereby due to gravity water reaches to home outlets.

The overhead tank can however be eliminated if water is supplied directly from underground tank to kitchen toilet outlets, there comes the need of pumps which can give uninterrupted supply of water with required pressure to outlets so that when one opens the tap he gets continuous supply of water. Such pumps are called hydro-pneumatic system.

Such pumps consists of small steel tank with water on one side and air on another separated by a rubber membrane. As the pump starts it supplies water to the wet side thus causing rubber membrane to expand and air compresses on other side thus causing extra pressure on wet side which is connected to water supply line. So as one opens the tap , gets the required quantity of water. This causes the pressure to drop and the pump is automatically switched on again thereby maintaining the pressure of water and at same time supplying the water to outlets.

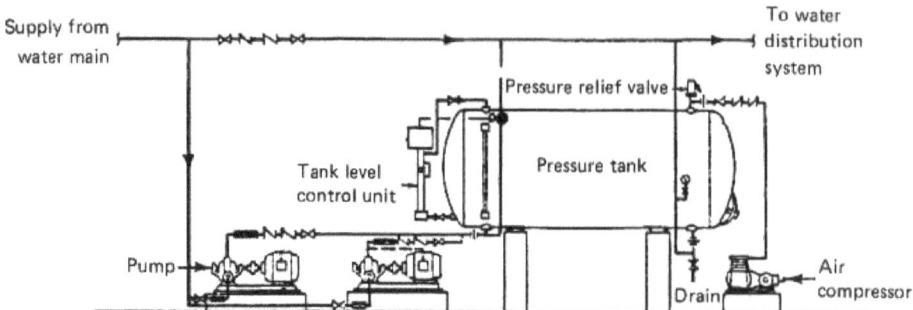

The advantage of such plumbing system in buildings is that requirement of overhead tanks is eliminated. Secondly these pumps are designed to get the required equal pressure to all floors, unlike in traditional way water flows from underground tanks to overhead tanks and the upper floors gets water with less pressure and lower & ground floor gets high pressure due to gravity and more height.

This gives energy conservation also because it eliminates the need of supply of water ten or twenty floors to overhead tank and then supply by gravity to all floors.

Continuous power supply backed up with generator is required to operate this system efficiently, else if no power no supply of water.

Pumps are of Two Types:

- Submersible pumps, and

- Open type Pumps

Submersible pumps are used inside the water and require very less maintenance. Both types can be used for traditional as well as hydro-pneumatic system.

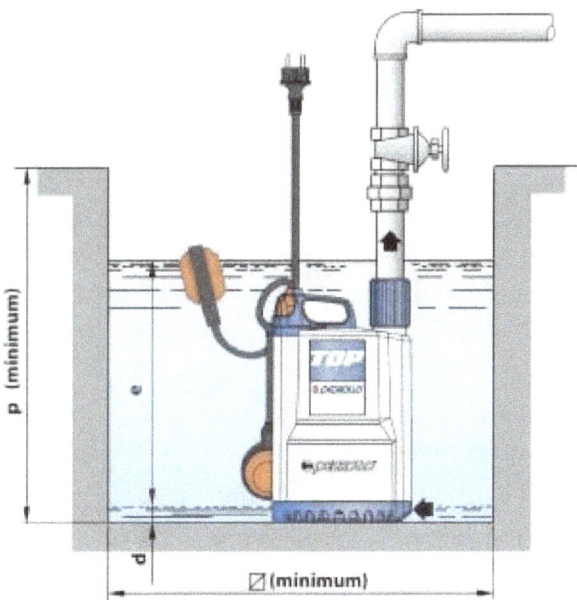

Types of Drainage Systems in Buildings

1. Waste water is from showers, basins, kitchen sinks, washing machines, and the like. This is also called grey water. Normally a minimum of 75 mm dia. pipes are used for drainage of waste water.

2. Soil water or sewage is from WCs and urinals. This is also called black water. Minimum of 100 mm diameter pipes are used for waste water. When run horizontally, soil water pipes should be run at a steeper slope, such as 1:40, as they have solids. These can be of cast iron or of PVC.

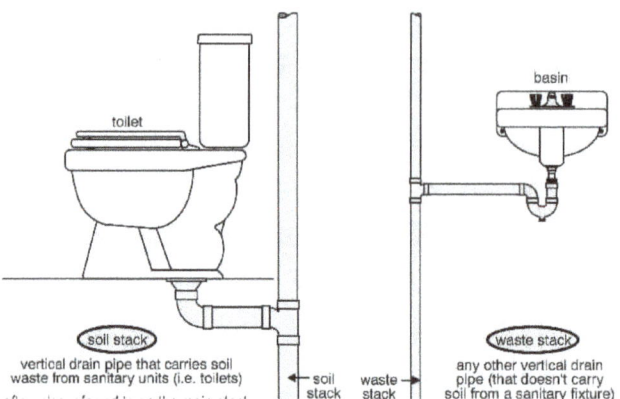

A grease trap should be used when draining waste from kitchens, grease should not be allowed to enter the normal drainage system. A grease trap is nothing but a small inspection chamber. The grease floats, and should be removed manually on a daily basis. The inlets and outlets into this chamber should be designed in a way that minimizes disturbance of the floating grease layer.

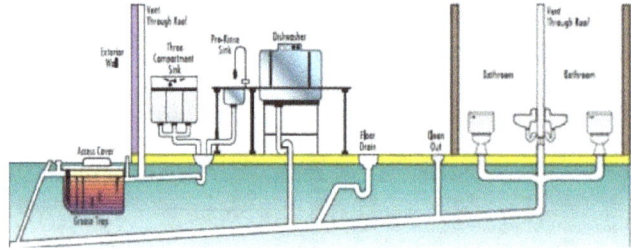

Stone ware (ceramic) pipes are used when soil and waste water is to be transported in external soil. An inspection chamber is used to clean blockage in the line and change direction of pipes. Inspection chamber is a short version of manholes which are used on the streets.

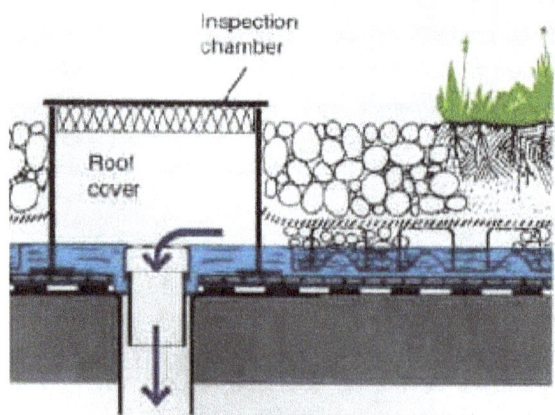

If municipal government drainage is not available on a small project, or exterior of city, provide a septic tank and a soak pit. A septic tank is a rectangular underground tank with compartments. It is always full of sewage that can be removed manually. The less water put into a septic tank, the better it will function. The effluent that flows out of this, which is about 70% purified, is then put into a soak pit. A soak pit is a cylindrical tank with porous brick walls surrounded by a layer of gravel.

A soak pit should not be placed near any occupied structure, water body, or water supply pipe. It also cannot be used where the water table is high, as groundwater will then enter and flood the pit through the porous walls.

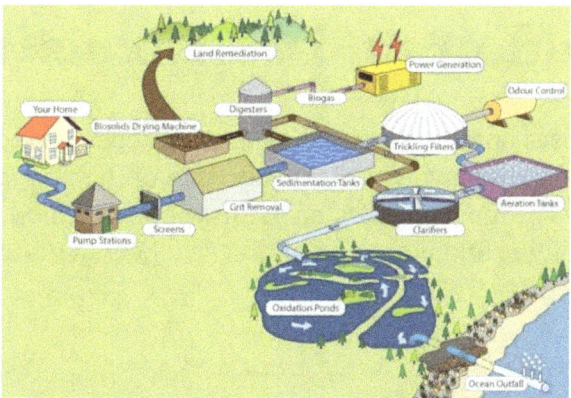

An overall view of complete sewage treatment of a complete city

A sewage treatment plant is recommended for the disposal of large amounts of sewage. This is a plant that will process sewage and produce sludge and (relatively) clean water from it. This water may then be used for landscaping, or even perhaps for HVAC cooling towers – not for drinking or washing.

References

- Structural-system, technology: britannica.com, Retrieved 19 March 2018

- Ventilation-systems-exhaust, how-it-works: hometips.com, Retrieved 29 April 2018

- Roof-construction, how-it-works: hometips.com, Retrieved 31 March 2018

- What-is-roof-decking: roofingsouthwest.com, Retrieved 22 March 2018

- Types-of-ceiling-construction-14204: theconstructor.org, Retrieved 29 May 2018

- The-three-major-types-of-plumbing-systems: dmsplumbinglasvegas.com, Retrieved 19 May 2018

- Types-of-plumbing-drainage-systems-buildings-13117: theconstructor.org, Retrieved 11 July 2018

Modular Building and Construction

Modular buildings are houses or structures that consists of repeated modules or sections. Its construction is a unique process that involves the installation of the prefabricated sections at the intended site of the building. The topics elaborated in this chapter on modular construction, modular building, prefabricated building and portable building will help in developing a better perspective of modular building and construction.

Modular Construction

Modular construction is a process in which a building is constructed off-site, under controlled plant conditions, using the same materials and designing to the same codes and standards as conventionally built facilities – but in about half the time.

Modular construction' is a term used to describe the use of factory-produced pre-engineered building units that are delivered to site and assembled as large volumetric components or as substantial elements of a building. The modular units may form complete rooms, parts of rooms, or separate highly serviced units such as toilets or lifts. The collection of discrete modular units usually forms a self-supporting structure in its own right or, for tall buildings, may rely on an independent structural framework.

Modular hospital building during installation of open sided modules

The main sectors of application of modular construction are:

- Private housing
- Social housing
- Apartments and mixed use buildings
- Educational sector and student residences

- Key worker accommodation and sheltered housing

- Public sector buildings, such as prisons and MoD buildings

- Health sector buildings

- Hotels.

Thousands of modules are manufactured annually in the UK. The largest markets for modular construction are in student residences, military accommodation and hotels, but the health sector is significant as it requires highly complex services and medical installations that can be commissioned and tested off-site.

Attributes of Modular Construction

Modular units used in multi storey social housing project in London

The use of modular and other lightweight forms of building construction is increasing. The benefits of modular construction, relative to more traditional methods, include:

- Economy of scale through repetitive manufacture

- Rapid installation on site (6-8 units per day)

- High level of quality control in factory production

- Low selfweight leading to foundation savings

- Suitable for projects with site constraints and where methods of working require more off-site manufacture

- Limited disruption in the vicinity of the construction site

- Useful in building renovation projects, such as roof top extensions

- Excellent acoustic insulation due to double layer construction

- Adaptable for future extensions, and ability to be dismantled easily and moved if required

- Robustness can be achieved by attaching the units together at their corners

- Stability of tall buildings can be provided by a braced steel core.

Modular construction is most commonly associated with cellular type buildings such as student residences or key worker accommodation. For these applications it has the following features:

- Suitable for buildings with multiple repeated units

- Size of units is limited by transport (3.6m x 8m is typical)

- Open sided units can be created (by changing the floor orientation)

- Modules are stacked with usually no independent structure

- Self weight of 1.5 to 2 kN/m²

- 4 to 10 storeys (6 is usually the optimum)

- Fire resistance of 30 to 60 minutes provided

- Acoustic insulation is provided through double layer walls and floors.

The following table describes the various structural elements used in walls and floors of modules.

Walls	Floors
Walls of modules comprise C sections of 75 to 150 mm depth.	Floors of modules comprise C sections of 150 to 250 mm depth.
Longitudinal walls are usually load-bearing and the end walls provide for stability.	Ceiling is manufactured as a wall panel.
Open-sided modules can be created by longitudinal floor and ceiling joists – end walls become load bearing.	Open-sided modules use deeper floor joists or lattice joists of 250 to 400 mm depth.
Stability provided by cross-flats or diaphragm action of boarding.	Corridor zone can be used to provide in-plane bracing in long buildings.
Double skin walls provide excellent acoustic insulation.	Double skin floor and ceiling provides excellent acoustic insulation. Mineral wool may be required between the joists.
Construction of walls and floors in modular units	

Types of Modules

The following types of modules may be used in the design of wbuildings using either fully modular construction or mixed forms of steel construction:

- 4-sided modules

- Partially open-sided modules

- Open-sided (corner-supported) modules

- Modules supported by a primary structural frame

- Non-load bearing modules

- Mixed modules and planar floor cassettes

- Special stair or lift modules.

The structure of the modules consists mainly of light steel C sections that are cold rolled from strip steel to BS EN 10346. Additional corner posts in the form of square hollow sections are often used.

4-sided Modules

In this form of construction, modules are manufactured with four closed sides to create cellular type spaces designed to transfer the combined vertical load of the modules above and in-plane loads (due to wind action) through their longitudinal walls. The cellular space provided is limited by the transportation and installation requirements. Depending on location and exposure to wind action, the height of buildings in fully modular construction is in the range of 6 to 10 storeys.

Modules are manufactured from a series of 2D panels, beginning with the floor cassette, to which the four wall panels and ceiling panel are attached generally by screws. The walls transfer vertical loads and therefore the longitudinal walls of the upper module are designed to sit on the walls of the module below.

Additional steel angles may be introduced in the recessed corners of the modules for lifting and for improved stability. Module to-module connections are usually in the form of plates that are bolted on site. Special lifting frames are used that allow the modules to be unhooked safely at height.

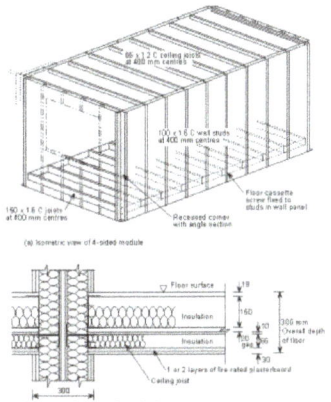

Details of 4 sided modules showing recessed corners with additional angle sections

Module being lifted in the factory

Modules can be manufactured with integral balconies and a range of cladding materials can be pre attached or installed on site. All walls are insulated, and are usually boarded externally for weather protection. Additional external insulation can be attached on site.

For low rise buildings, in plane bracing or diaphragm action of the board materials within the modules provides shear resistance, assisted by the module to module connections, which transfer the applied wind forces to the group of modules.

For buildings of 6 to 10 storeys height, a vertical bracing system is often located around an access core, and assisted by horizontal bracing in the corridor floor between the modules. For taller buildings, a steel podium frame may be provided on which the modules are stacked and supplemented by a concrete or steel core.

The maximum height of a group of modules is dependent on the stability provided under wind action. Various cases are presented in the table for scheme design (based on wind loading in the Midlands of England).

Form of modular construction	Bracing requirements	Limit on size in concept design	
		Typical max. number of storeys	Min. number of modules in a group
Single line of modules	No additional bracing	3	5
	With additional bracing in gables	5	8
	With additional stabilising core	7	No limit
Double line of modules with central corridor	No additional bracing	6	2 x 8
With additional bracing in gables		8	2 x 10
	With additional stabilising core	10 - 12	No limit
Typical building height depending on the stabilising system using 4 sided modules general guidance for scheme design			

Partially Open-sided Modules

4 sided modules can be designed with partially open sides by the introduction of corner and intermediate posts and by using a stiff continuous edge beam in the floor cassette. The maximum width of opening is limited by the bending resistance and stiffness of the edge member in the floor cassette. Additional intermediate posts are usually square hollow sections(SHS), so that they can fit within the wall width.

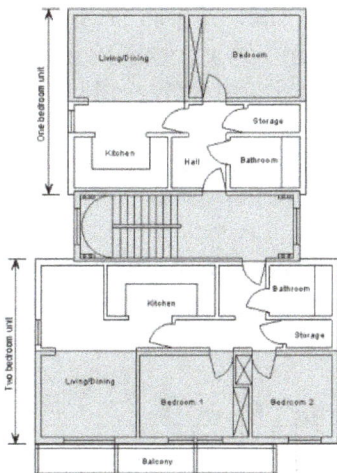

Layout of apartments using partially open sided modules – alternate modules are shaded

Two modules can be placed together to create wider spaces. The compression resistance of the corner or internal posts controls the maximum height of the building, but 6 to 10 storeys can be achieved, as for fully modular construction.

Long modules can also be designed to include an integral corridor, as shown below. The length of the module may be limited by transport and site access but a length of up to 12m is normally practical. Use of modules with integral corridors can improve the speed of construction by avoiding weather tightness problems during installation and finishing work.

Partially open ended module used in Barling Court, London

Long module with an integral corridor

The form of construction is similar to that of 4 sided modules, except for the use of additional posts, generally in the form of 70 x 70 to 100 x 100 SHS members. Balconies or other components can be attached to the corner or internal posts. Overall stability is provided by additional bracing located in the walls of the modules.

Stability of the modules is affected by their partially open sides; additional temporary bracing during lifting and installation may be necessary. A separate bracing system may also be required, as the partially open-sided modules may not possess sufficient shear resistance in certain applications. A typical building form in which larger apartments are created using partially open sided units is shown right.

Open Sided (Corner-supported) Modules

Primary steel frame used in a fully open sided module

Modules may be designed to provide fully open sides by transfer of loads through the longitudinal edge beams to the corner posts. The framework of the module is often in the form of hot rolled steel members, such as Square Hollow Section (SHS) columns and Parallel Flange Channel (PFC) edge beams, that are bolted together.

A shallower parallel flange channel (PFC) section may be used to support the ceiling, but in all cases, the combined depth of the edge beams is greater than for 4 sided modules. Modules can be placed side by side to create larger open plan spaces, as required in hospitals and schools, etc.

The stability of the building generally relies on a separate bracing system in the form of X bracing in the separating walls. For this reason, fully open ended modules are not often used for buildings more than three storeys high. Where used, infill walls and partitions within the modules are non load bearing, except where walls connected to the columns provide in plane bracing. The corner posts provide the compression resistance and are typically 100 x 100 SHS members. The edge beams may be connected to these posts by fin plates, which provide nominal bending resistance. End plates and Hollo-bolts to the SHS members may also be used. The corner posts possess sufficient compression resistance for use in buildings at least up to 10 storeys.

As open sided modules are only stable on their own for one or two storeys, additional vertical and horizontal bracing is usually introduced. In plane forces can be transferred by suitable connections at the corners of the modules.

An open ended module is a variant of a 4 sided module in which a rigid end frame is provided, usually consisting of welded or rigidly connected Rectangular Hollow Sections (RHS). The rigid end frames are manufactured as part of the module or can be assembled as separate components.

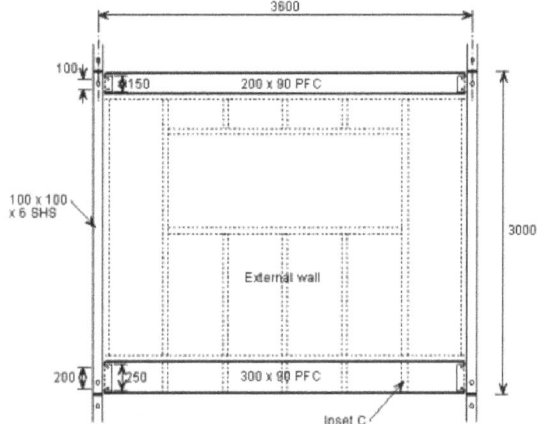

Structural frame of a corner supported module – end view

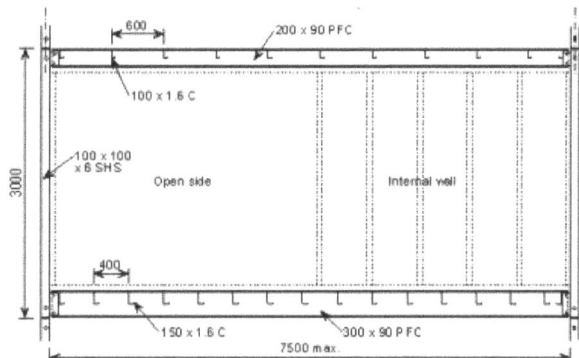

Longitudinal edge beams of a corner supported module

A steel external framework comprising walkways or balconies may be also designed to provide stability. Modules using hot rolled steel framework can be designed to support concrete floors for use in medical and other applications, where strict control of vibrations is required.

Mixed Modules and Floor Cassettes

In this 'hybrid' or mixed form of construction, long modules may be stacked to form a load-bearing serviced core and floor cassettes span between the modules and load-bearing walls. The modules are constructed in a similar way to that described for open-sided modules, but the loading applied to the side of the modules is significantly higher. Therefore, this mixed modular and panel form of construction is limited to buildings of 4 to 6 storey height. It is typically used in residential buildings, particularly of terraced form, comprising modular 'cores' for stairs, and highly serviced areas. The modules are arranged in a 'spine' through the building and the floors are attached to it. An example of this hybrid form of construction is shown.

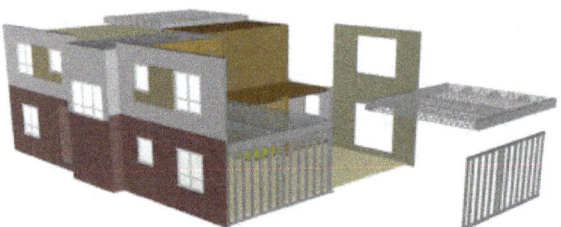

'Hybrid' modular and panel building showing planar floors and walls attached to the modules

Load-bearing bathroom modules were used to support the floor cassettes at Lillie Road.

Modules Supported by a Primary Structure

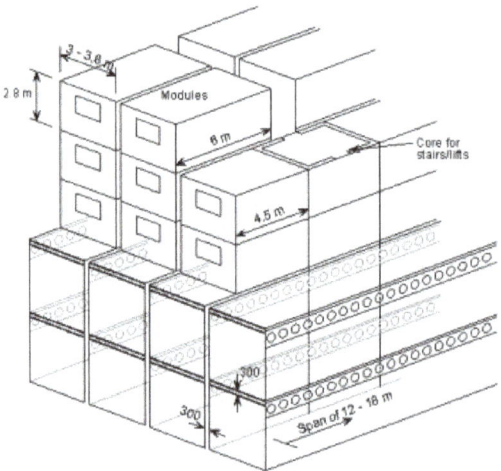

Modules supported by long spanning cellular beams to create open plan space at the lower levels

Modular units may be designed to be supported by a primary structure at a podium or platform level. In this case, the supporting columns are positioned at a multiple of the width of the modules (normally 2 or 3 modules). The beams are designed to support the combined loads from the modules above (normally a maximum of 4 6 storeys).

The supporting structure is designed conventionally as a steel framework with beams and columns that align with multiples of the module width, and provides open plan space at ground floor and below ground levels. This form of construction is very suitable for mixed retail, commercial and residential developments, especially for residential units above commercial areas or car parking, etc, particularly in urban projects.

Modules can be set back from the facade line. An example of a mixed development in Manchester is shown. The ground floor and below- ground car parking is a conventional composite structure.

Typical podium structure in which seven storeys of residential units are supported on a composite structure frame below)

Where the 4 sided modules are designed to be supported by steel or composite beams and the typical line load per supported floor is 15kN/m, columns are placed at 6 to 8m spacing. A column spacing of 7.2m is suitable for below ground car parking. The depth of the podium type structure is 800 to 1000mm, and spans of 10 to 18m can be created below the podium, which are suitable for commercial applications and car parking.

The podium structure is generally braced to resist wind loads and a separate braced core is often used to stabilise the group of modules above the podium level. The module design is similar to that described for 4 sided modules. Wind loadscan be transferred horizontally through the corridor floors.

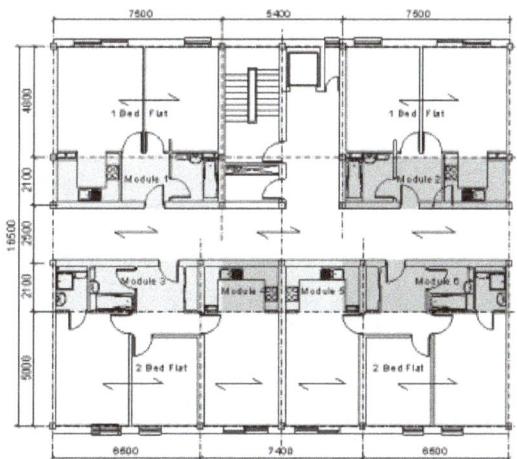

Mixed use of modules and long spanning floor with a primary steel frame

Alternatively, non load bearing modules can be supported by a primary frame, and are installed as the construction proceeds. Modules can be disassembled in the future to leave the floor cassette supported by the beams. An example of the mixed use of modules and primary steel frame is shown in below figure. The modules are shown shaded and floor spans indicated.

An external steel structure, consisting of a façade structure that acts to stabilise the building, may also be used. Modules are placed internally within the braced steel frame, as shown in the MoHo project in Manchester.

Installation of modules behind external steel framework at MoHo, Manchester

Other Types of Modules

Various forms of other modular components have been used in major building projects.

Stair Module

Modular stairs may be designed as fully modular units and generally comprise landings and half landings with two flights of stairs. The landings and half landings are supported by longitudinal walls with additional angles or SHS members to provide local strengthening, if necessary. The stair modules rely for their stability on a base and top, which leads to use of a false landing. The open top and base of the wall may be strengthened by a T, L or similar members to transfer out of plane loads to the landing. SHS posts and bracing can be introduced in the walls to provide for overall stability.

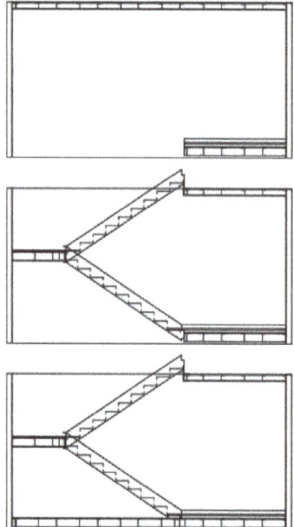

Detail of light steel modular stair system showing use of a 'false' landing to modules

Non Load Bearing Modules

Non load bearing modules are of similar form to fully modular units, but are not designed to resist external loads, other than their own weight and the forces during lifting. They are used as toilet/ bathroom units, plant rooms or other serviced units and are supported directly on a floor or by a separate structure. The walls and floor of these 'pods' are relatively thin (typically <100mm). The units are designed to be installed either as the construction proceeds or slid into place on the completed floor.

Compatibility of the floor depth in the module and in the floor elsewhere is achieved by one of four methods:

- Designing the depth of the floor of the module to be the same as the raised floor or acoustic layer elsewhere.

- Placing the module in a recess in the floor of the main structure.

- Designing the module without a floor (possible in small modules in which fitments are attached to the walls).

- Designing the modules to be supported on the bottom flange of Slimflor beams.

Balconies and Atria

Atrium created between modulesm

Balconies may be attached to modules in various ways:

- Balconies supported by a self standing steel structure that is ground supported.

- Balconies attached between adjacent modules.

- Balconies that are attached to corner posts in the modules.

- Integrated balconies within an open sided module.

Atria may be created by attaching a lightweight steel roof to the upper modules or by by spanning the roof between the modules as shown.

Balcony attachments to external structure (MoHo, Manchester)

Key Technical Issues

The following general design issues are reviewed below:

- Dimensional planning
- Stability and structural integrity
- Service interfaces
- Acoustic performance
- Fire safety.

Dimensional Planning

The factors that influence the dimensional planning of modular systems in general building design may be summarised as:

- Cladding requirements, including alignment with external dimensions of cladding
- Planning grid for internal fit out, such as kitchens
- Transportation requirements, including access to the site
- Building form, as influenced by its functionality
- Repeatability in modular manufacture.

Cladding

Brickwork design is based on a standard unit of 225mm width and 75mm depth. Therefore, it may be important to design a floor depth to a multiple of 75mm in order to avoid non standard coursing of bricks.

Other types of cladding, such as clay tiles or metallic finishes, have their own dimensional requirements, but generally they can be designed and manufactured to fit with window dimensions etc. Many types of lightweight cladding can be pre attached to the modules, but it is generally necessary to install a cover piece over the joints between the modules on site, to cater for geometrical tolerances and misalignments.

Standardisation of Planning Grid

Standardisation of the planning grid is important at the scheme design stage, as the planning grid will be controlled by other building components and fitments. A dimensional unit of 300mm may be adopted as standard for vertical and horizontal dimensions, reducing to 150mm as a second level for vertical dimensions. External walls are detailed according to the type of cladding, but a 300mm total wall width may be adopted as a guide for most cladding materials. The actual width will vary between 200mm for insulated render and board materials to 320mm for brickwork.

Typical dimensions for planning in modular construction are presented in the table.

Application	Internal wall height (mm)	Internal module width (mm)	Internal module length (m)	Ceiling-floor zone (typical) (mm)
Study bed-rooms	2400	2500 – 2700	5.4 to 6	300
Apartments	2400	3300 – 3600	6 to 9	450
Hotels	2400 – 2700	3300 – 3600	5.4 to 7.5	450
Schools	2700 – 3000	3000 – 3600 open-sided	9 to 12	600
Offices	2700 – 3000	3000 – 3600	6 to 12	600 – 750
Health sector	2700 – 3000	3000 – 3600 open-sided	9 to 12	600 – 750
Typical dimensions for planning in modular construction				

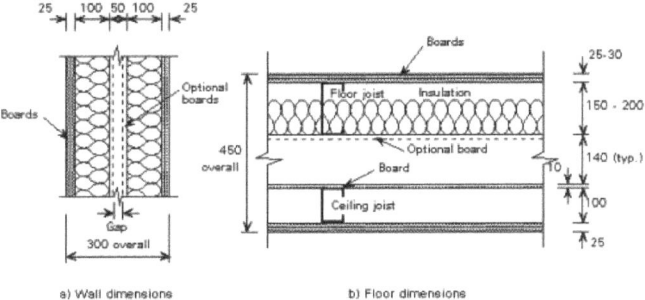

Typical wall and floor/ceiling dimensions

Transportation

Guidance on transportation on major roads is given by the Road Haulage Association, based on the Road Vehicles (Construction and Use) Regulations.

The following basic requirements for transportation should be considered when designing the sizes of modular units:

- Modules exceeding 2.9m external width require 2 days notice to the police

- Modules exceeding 3.5m width require a driver's mate and 2 days police notice

- Modules exceeding 4.3m width require additional speed restrictions and may require police escort.

Stricter limits may be required for local roads, particularly in urban areas. In all cases, the maximum height of the load is 4.95m for motorway bridges. Standard container vehicles can deliver one large or two smaller units.

Internal Walls

Internal walls comprising the walls of adjacent modules may be designed for a standard 300mm face face overall width, incorporating the sheathing boards, internal plasterboards and insulation between the C sections. The gap between the walls is a variable, depending on the number and thickness of boards and size of the wall studs.

Floor Zone

Floors and ceilings in modular construction are deeper than in more traditional construction. The three structural cases of side supported (4-sided modules), corner supported (open sided) and frame supported modules require different overall ceiling floor dimensions for planning purposes, as follows:

- Continuously supported or 4-sided modules: 300 or 450mm

- Corner supported or open-sided modules: 450 to 600mm

- Frame supported modules: 750 to 900mm.

In most cases, 450mm may be adopted as a standard for the floor-to-ceiling dimension, although many systems provide shallower depths. For corner supported modules, a standard overall floor and ceiling depth of 600mm may be used. The gap between the floor and ceiling is a variable depending on the number of boards and the joist size.

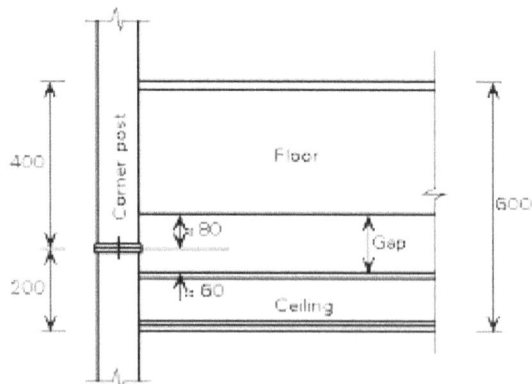

Detail of corner supported module

Overall stability is provided by the modules themselves, or by an external structure. The load path is through the walls of the 3-D units, and so removal of this load path means that the walls should be designed to either:

- Span horizontally over a damaged area by acting as a deep beam ,or

- Be supported by tie forces to the adjacent units.

Stability and Structural Integrity

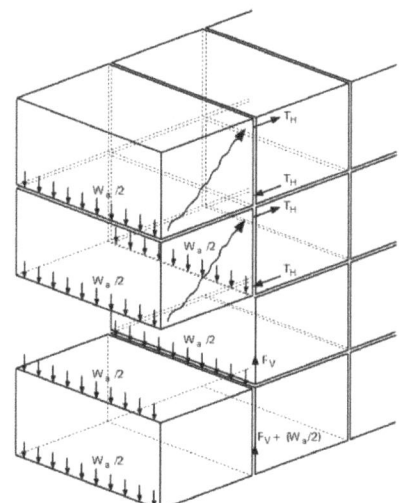

Tying forces in modular construction subject to loss of one module by fire or explosion

The latter means that the units should be tied both horizontally and vertically. Robustness is provided by the ties between the modules with a normally assumed minimum tying force equivalent to half the loaded weight of the module (minimum value of 30kN).

Service Interfaces

The installation of electrical, plumbing and heating services in modular buildings can be largely carried out in the factory with final connections made on-site. In traditional construction, such activities are labour intensive on-site and are often on the critical path, so that any difficulties can cause delays. Service strategies that have been used in modular buildings include:

- Use of communal spaces for distribution of services

- Use of the floor or ceiling zone within each module for service distribution

- Installation of services within each module in the factory with site work involving only connection of modules

- Drainage connections of modules connected to vertical risers in the corner of the modules

- Wet areas are connected back to back to concentrate service zones.

A vertical service duct is usually incorporated in the corner of each unit to accommodate the vertical drainage and pipework. The services in each module are installed in the factory and terminate in the vertical duct. Access to the service duct is generally only possible from circulation areas outside the modular unit.

The horizontal distribution of services between modules varies, depending on the building type. For most types of residential buildings and hotels, the corridor ceiling and floor voids act as service zones.

Vertical drainage stacks are also installed in the factory and a removable floor panel is provided to allow the final connection to the drains installed in the ground on-site. This requires a high degree of accuracy in setting out service inlets on-site.

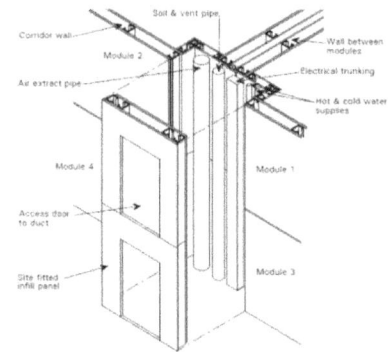

Typical service duct in a modular hotel building

Acoustic Performance

Modular construction provides a high level of acoustic separation because each module has separate floor, ceiling and wall elements, which prevents direct transfer of sound along the members.

Modular unit manufacturers use various methods to further improve sound reduction between units – two overlapping layers of plasterboard fixed inside each module, oriented strand board (OSB) or plasterboard fixed as external sheeting or quilt insulation between steel members.

Special care needs to be taken around openings for service pipes or other penetrations, because sound attenuation is particularly affected by air pathways between spaces. Electrical sockets penetrate the plasterboard layer, so they should be carefully insulated using quilt at their rear.

Fire Safety

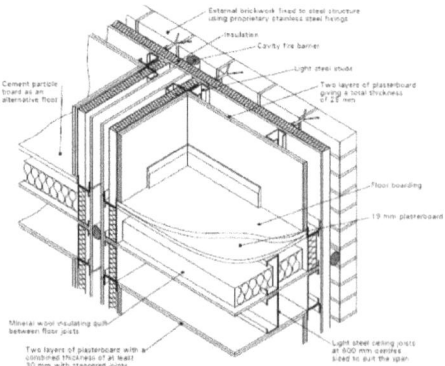

Compartment floor at junction with external wall and compartment wall

Fire safety is related to provision of adequate means of escape, to ensuring structural integrity, and controlling spread of fire across compartment boundaries.

Modular construction generally achieves these requirements by the use of fire resistant plasterboard conforming to BS EN 520, Type F. Alternative materials, such as cement particle board and gypsum fibre board may also be used in combination with plasterboard as the facing layer.

Each module is lined internally with one or two layers of fire resistant plasterboard as follows:

- For walls: 30 minutes fire resistance is achieved by a single layer of 12.5mm fire resistant plasterboard on each face of a steel stud wall

- For walls: 60 minutes fire resistance is achieved by one layer of 12.5mm fire resistant plasterboard on a layer of 12.5 mm wallboard with staggered joints on each face of a steel stud wall

- For floors: 30 minutes fire resistance is achieved with 18mm tongue and groove boarding on light steel joists and 12.5mm fire resistant plasterboard beneath with joints taped and filled

- For floors: 60 minutes fire resistance is achieved with at least 18mm T&G board floor finish and one layer of 12.5mm fire resistant plasterboard on a layer of 12.5mm wallboard with staggered joints beneath the steel joists.

In residential construction, each dwelling usually forms a separate fire compartment. All walls and floors that provide a separating function between compartments require 60 minutes fire resistance. In hotels and other residential buildings, each bedroom may form its own compartment.

In general, a compartment floor will also act as a separating floor for acoustic purposes, as the same measures will also achieve excellent acoustic insulation between rooms.

The inherent separation between modules provides an effective barrier to spread of fire. Means of escape should be considered early in the scheme design in order to ensure that the module design and layout can satisfy these requirements. Cavity barriers are required within the cavity in the external wall between the module and the cladding at intersections with compartment walls. They are also required horizontally at junctions with floors and roof, and vertically at a maximum lateral spacing of 20m (or 10m where the material exposed to the cavity is not Class 0 or 1 as defined by Approved Document B). Fire stops must be provided around any penetrations through fire resisting walls.

Sustainability

The concept of using sustainability indicators is becoming accepted as part of the environmental assessment of building construction. For modular construction, it is appropriate to include whole life measures, such as potential re-use, or re-location which are not properly reflected in conventional measures of sustainability.

The sustainability indicators relevant to modular construction are listed below. Comments on how modular construction contributes to these indicators are given against each indicator.

Energy Efficiency

Sustainability indicator	Comment on modular construction
Minimise energy in use	Good level of thermal insulation
	Efficient heating and cooling systems
	Control systems for energy saving provided

Energy saving measures	Efficient manufacture
Minimise CO_2 production from fossil fuels	Efficient operation and thermal insulation
	Efficient use of materials
Minimise embodied carbon in materials	Factory controlled operation
	100% recyclable
	High strength to weight ratio
	Ease of deconstruction and re-use
Minimising transport	

Sustainability indicator	Comment on modular construction
Suitable site location	Depends on public transport and adjacent public amenities (site specific)
Minimise transport impact	Raw materials delivered in bulk to factory
	Modules delivered to site fitted out
	Reduced deliveries to site
	Fewer personnel on site
Minimising pollution	

Sustainability indicator	Comment on modular construction
No use of ozone-depleting substances	Insulation materials selected to suit client needs
Minimise waste creation and disposal	Efficient use of materials in factory
	Minimum / zero waste on site
	Recycling of scrap metal
	Re-use of modules or components
Maximise waste recycling ratio	Recycled steel used in manufacture
	Modules can be re-used
Minimise nuisance in construction	Noise, vibration and dust reduced
	Fast construction process
	Less waste disposal
	Fewer site deliveries
Efficient materials and resource use	

Sustainability indicator	Comment on modular construction
Efficient use of materials	Steel has high strength to weight ratio
	Efficient design in materials use
	Long design life
Ability to be recycled or re-used	High proportion can be recycled
	Modules can be re-used
Low maintenance	Few 'call backs' due to quality of production
	Accessibility for maintenance is easier
Provision for future adaptability	Ability to extend / modify building
Health and well-being	

Sustainability indicator	Comment on modular construction
Maximise site safety	Manufacturing process is safe
	Safer site operations in modular construction
Considerate construction	Speed of construction on site
	Minimum noise, disruption etc.
Good acoustic insulation	Good insulation between modules
	Facades insulated against external noise
Adequate day lighting	Large windows can be provided in modules
Worker welfare	Safe and clean manufacture and construction
	Good operational conditions

Procurement

The typical procurement process for modular buildings requires that the client, designer and manufacturer work together at all phases of the project to maximise the benefits of the off-site process and manufacturing efficiency.

Decision-making Process

The decision-making process for modular construction differs from more traditional methods of construction because of:

- The close involvement of the client in assessing the business-related benefits provided by the method of construction.

- The direct involvement of the manufacturer in the design, costing and logistics.

- The need to make key decisions early in the procurement process

- The important environmental and site-related benefits that can be achieved

- The effect of transportation logistics on costs and module sizes.

Since the benefits of modular construction are realised through pre-fabrication, the initial design phase, including the space planning and subsequent detailed design, service integration, and co-ordination, are critical.

Procurement Process

In modular construction, the procurement process involves the specialist manufacturers. There are several ways of procuring modular buildings:

- Traditional, in which an architect provides the design co-ordination and the general contractor provides the construction co-ordination. The module manufacturer acts as a specialist sub-contractor.

- 'Design and Build' process, in which the module manufacturer provides the detailed design and construction responsibilities. In this case, the client's architect may carry out some of the outline design, and may be novated by the client to work for the contractor.

A Design and Build contract is often used for modular construction. In such cases, the role of the client's architect will depend on the particular procurement process. Two methods of specification by the architect are most commonly used:

- The architect may specify the manufacturer who will undertake the detailed design work. This will enable the parties to work together from inception to completion.

- Alternatively, the architect may draft a performance specification for the work. This is then used as a basis for tendering, either through a main contractor or directly to the modular specialists.

It should be recognised that each manufacturer undertakes the construction of their modular units differently and they will be prepared to offer advice and provide drawings at the concept stage.

Importantly, the 'lead-in' time required for prototype, design and manufacture of bespoke module units should be considered, although detailed design of the modular units can be carried out in parallel with other design activities. If the module configuration is repeated from other projects, then design time is much reduced.

Typical Details

Connections

Guidance on the design and detailing of the most common connection types is given in BS EN 1993-1-8. Manufacturers use the method which best suits their manufacturing process and for which appropriate test data are available.

Structural connections between modules are required for integrity and robustness but details vary depending on the form of the module and the particular application. Floor boarding, plasterboard and sheathing boards are attached using self drilling, self-tapping screws. Manufacturers of light steel framed modules have prepared their own details of horizontal attachments that satisfy robustness requirements.

Attachment Points

Attachments between modules are made in both horizontal and vertical directions, primarily to transfer in plane forces, but also for structured integrity.

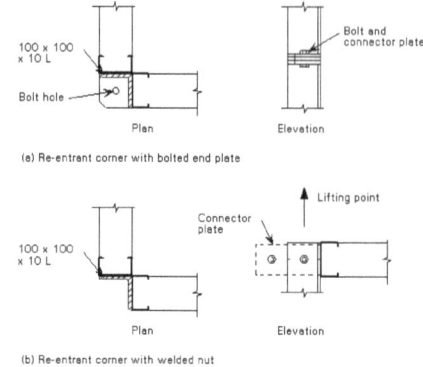

Corner posts using hot rolled steel angles

SHS provide the highest compressive resistance and may be used as the corner posts for open sided modules. However, although these sections are compact, their connections can be more complex. A welded fin plate to which the edge beams are bolted is shown. Access holes in the SHS allow bolts to be inserted through end plates to provide for vertical and horizontal attachments.

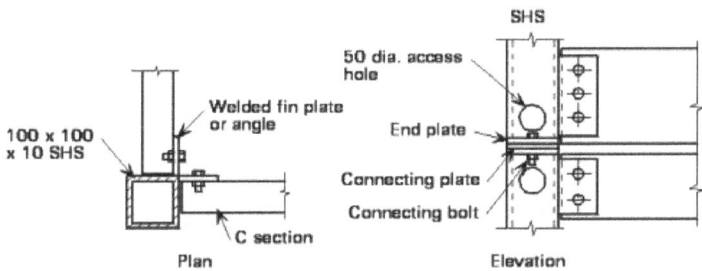

Corner post using or special sections

Facades and Interfaces

Various interfaces between modular units and other components in the building may not be under the control of the modular manufacturer. The responsibility for design and coordination usually lies with the building designer.

Foundation Interfaces

A variety of foundations can be used, including strip, trench-fill, pad and piled foundations. Strip or trench-fill foundations are most common.

Modular units are lightweight and therefore foundations may be smaller than in traditional construction. Nevertheless, the cladding options and building height may dictate the foundation design. With strips, rafts or ground beams, the modular units can be designed to be continuously supported around the perimeter of each unit.

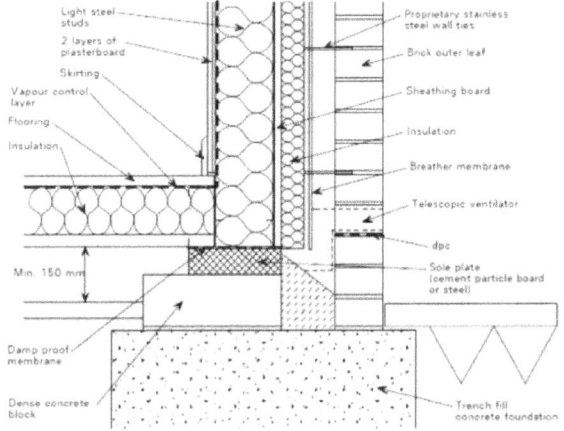

Typical trench-fill foundation detail for masonry cladding

The levelling of the foundations or ground beams is crucial to the subsequent installation and alignment of the modular units. The modular manufacturers have developed their own proprietary locating and fixing mechanisms to aid the positioning of units on the foundations.

Wall Cladding Interfaces

Claddings for modular buildings can be self supporting vertically and only supported laterally by the units. Alternatively, they can be supported entirely by the modular structure.

Two generic systems of facade construction may be considered:

- Cladding that is placed entirely on-site using conventional techniques.

- Cladding that is completely or partially attached in the factory; infill pieces or secondary cladding may be fixed on-site.

Cavity barriers must also be incorporated into any cavity that occurs between the external cladding and the modular structure. These must resist the spread of smoke and flame and are required between all separate dwellings or fire compartments. Mineral wool is generally used.

Roofing Interfaces

Roofing materials for modular buildings generally comprise tiles supported on battens, or roof sheeting on purlins. Modern roofs may comprise tiles supported on roof sheeting or structural liner trays. Flat roofs can also be constructed with a variety of weatherproof finishes. Insulation in the line of the roof pitch is used where a 'warm roof' is created. However, in most cases, the roof space is 'cold', and insulation is placed directly on the upper surface of the modular units.

Roofs are generally designed as separate structures that are supported either continuously by the internal walls of the modular units, or as free spanning roofs between the outer walls. Roofs may also be designed as modular units for habitable space, and ease of installation, especially in taller buildings. However, conventional trussed rafter or purlin roofs are mostly used.

Roofs are designed to support the weight of the roof covering, snow loads, services and tanks stored on the roof space, and occupancy loads from habitable use. The interface between the roof and the modular units is designed to resist both compression and tension due to wind uplift. In some cases, the roof can be designed to be detachable so that the building can be extended later. Shallow pitch roofs can be designed to be supported directly by the modular units and are easily dismantled.

Modular Building

'Modular Building Construction' is a term used to describe the use of factory-produced pre-engineered building units that are delivered to site in Modules. The modular units may form complete rooms, or parts of rooms.

Modular buildings are generally made into six sided boxes constructed in a factory, then delivered to site and using a crane, the modules are set onto the building's foundation and joined together to make a single building. The modules can be placed side-by-side, end-to-end, or stacked, allowing a wide variety of configurations and styles in the building layout. Typically in South Africa Modular Buildings are referred to as "Parkhomes".

Types of Modular Building

Modular Trailers

Outside of modular homes, modular trailers are probably the type of modular building that people are most familiar with. These buildings have a lot of applications and are often used as a temporary space solution. What're the benefits of these modular trailers?

Well, these buildings are easily moved and there's are readily available to be rented or bought across the country. Unlike most of the some of the other types of modular buildings, these modular trailers are oftentimes just rented for months or years at a time. So oftentimes the trailer will be used by a number of different companies by the end of its life. Because these modular buildings are easily accessible and typically used as rentals, they are very popular on construction sites. For construction projects that last months at a time, these modular trailers provide a comfortable and temporary work environment for the construction foreman and engineers. Then, when the project is over, the construction company can return the building. These buildings are also commonly used as temporary classrooms for growing school systems that need class space before the beginning of the coming school year.

Containers

When looking at modular containers, they are structures that are typically used to contain a ship's cargo. However, recently these containers have been adapted to serve other purposes here on land. Typically, the containers can be customized to do a whole lot more than hold goods. There has been a shift recently where after these buildings serve their purpose as shipping containers, they are reused, fitted with windows, doors, HVAC units, etc. And made into an office environment. The biggest benefit of these containers is they are cheap, very portable, and easy to install on site. However, since the majority of the time these buildings are retrofitted and remodeled into modular buildings, they usually do not have a lot of flexibility in layout or their dimensions. Similar to the modular trailers, these container offices are often used at construction sites when they need to get office space quick and temporarily. Recently, these containers have been customized to be used for trendy storefronts, shops, or even stackable living spaces. But as with their application as offices, the design and space of the area can be oftentimes limited.

Panelized

Panel assembled office systems provide a different variety of benefits compared against other modular buildings. The biggest difference being the advantage of mass customization. Mass Customization, it states "the process of delivering wide-marketing goods and services that are modified to satisfy a specific customer need." In the case of Panel Built's panelized modular office systems, that is the main goal of our product. The entire office is comprised of a series of '4 x 8' panels, and the layout of the office system is at the full disposal of the end customer.

This office system can be installed at the job site in one of two fashions: pre-assembled or assembled on site. Like the previous two modular systems, the panelized offices can be delivered pre-built to the job site, only requiring anchoring and hook ups to be installed. Also, like the previous two, these preassembled versions will typically be limited to a rectangle shape. However, when assembling on site, the office system can take whatever shape the end customer needs and can even take advantage of pre-existing walls to complete the structure.

Office Complexes

The final form of modular building is going to be mostly seen in a more permanent type of modular structure. Office complexes will take a number of office modules and combine them to assemble a much larger, fully functioning office building. Just like all modular construction, these modules are fabricated in a manufacturing environment, except when these projects reach the job site, they function more similarly to a full-fledged construction site. Generally, the module will need a crane to lift the section from the ground and place it in the correct spot to assemble the building. Next, the workers will go through that module and do all things necessary to attach and hook up the piece. Then, the next piece is put in place, and these steps are repeated again and again until the building is finished.

Construction Process

Construction is offsite, using lean manufacturing techniques to prefabricate single or multi-story buildings in deliverable module sections. PMC buildings are manufactured in a controlled setting and can be constructed of wood, steel, or concrete. Modular components are typically constructed indoors on assembly lines. Modules' construction may take as little as ten days but more often one to three months. PMC modules can be integrated into site built projects or stand alone and can be delivered with MEP, fixtures and interior finishes.

The buildings are 60% to 90% completed offsite in a factory-controlled environment, and transported and assembled at the final building site. This can comprise the entire building or be components or subassemblies of larger structures. In many cases, modular contractors work with traditional general contractors to exploit the resources and advantages of each type of construction. Completed modules are transported to the building site and assembled by a crane. Placement of the modules may take from several hours to several days.

Permanent modular buildings are built to meet or exceed the same building codes and standards as site-built structures and the same architect-specified materials used in conventionally constructed buildings are used in modular construction projects. PMC can have as many stories as building codes allow. Unlike relocatable buildings, PMC structures are intended to remain in one location for the duration of their useful life.

Manufacturing Considerations

The entire process of modular construction places significance on the design stage. This is where practices such as Design for Manufacture and Assembly (DfMA) are used to ensure that assembly tolerances are controlled throughout manufacture and assembly on site. It is vital that there is enough allowance in the design to allow the assembly to take up any "slack" or misalignment of components. The use of advanced CAD systems, 3D printing and manufacturing control systems are important for modular construction to be successful. This is quite unlike on-site construction where the tradesman can often make the part to suit any particular installation.

| Walls attached to floor | Ceiling drywalled in spray booth | Roof set in place | Roof shingled and siding installed |

Pratt Modular Home "The   Interior Pratt Homes Wil-   Pratt Modular Homes "The   Ready for delivery to site
Willow" Tyler Texas   low Kitchen   Briar Ritz

Advantages

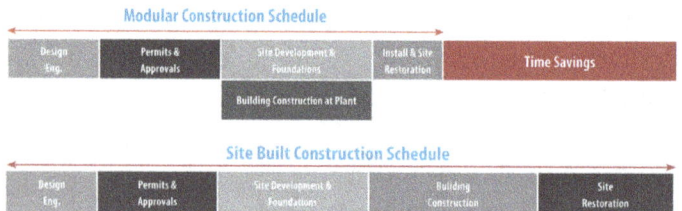

Simultaneous site development and building construction at the plant
reduces schedule by 30% to 50%

Modular buildings are argued to have advantages over conventional buildings, for a variety of reasons:

- Speed of construction/faster return on investment. Modular construction allows for the building and the site work to be completed simultaneously. According to some materials, this can reduce the overall completion schedule by as much as 50%. This also reduces labor, financing and supervision costs. To save even more time and money, nearly all design and engineering disciplines are part of the manufacturing process. Also unique to modular construction is the ability to simultaneously construct a building's floors, walls, ceilings, rafters, and roofs. During site-built construction, walls cannot be set until floors are in position, and ceilings and rafters cannot be added until walls are erected. On the other hand, with modular construction, walls, floors, ceilings, and rafters are all built at the same time, and then brought together in the same factory to form a building. This process can allow modular construction times of half that of conventional, stick-built construction.

- Indoor construction. Assembly is independent of weather, which can increase work efficiency and avoids damaged building material.

- Ability to service remote locations. Particularly in countries in which potential markets may be located far from industrial centers, such as Australia, there can be much higher costs to build a site-built house in a remote area or an area experiencing a construction boom such as mining towns. Modular buildings are also beneficial in providing medical and sanitary facilities where time, space, and money are an issue.

- Low waste. With the same plans being constantly built, the manufacturer has records of exactly what quantity of materials are needed for a given job. With the consistency, builders can design systems that use common lengths of lumber, wallboard, etc., cut items with maximum efficiency, or be able to order special lengths in bulk. While waste from a site-built dwelling may typically fill several large dumpsters, construction of a modular dwelling generates much less waste. According to the UK group WRAP, up to a 90% reduction in materials can be achieved through the use of modular construction. Materials minimized include: wood pallets, shrink wrap, cardboard, plasterboard, timber, concrete, bricks, and cement.

- Environmentally friendly construction process. Modular construction reduces waste and site disturbance compared to site-built structures. The controlled environment of the factory allows for more accurate construction while allowing the extra materials to be recycled in-house.

- Flexibility. One can continually add to a modular building, including creating high rises. When the needs change, modular buildings can be disassembled and the modules relocated or refurbished for their next use reducing the demand for raw materials and minimizing the amount of energy expended to create a building to meet the new need. In essence, the entire building can be recycled in some cases.

- Quality. Combining traditional building techniques, quality manufacturing and third-party agencies who offer random inspections, testing, and certification services for quality control, commercial modular buildings are built in strict accordance with appropriate local, state, and national regulations and codes. Due to the need to transport modules to the final site, each module must be built to independently withstand travel and installation requirements. Thus the final module-to-module assembly of independently durable components can yield a final product that is more durable than site-built structures. Modular buildings often use Structural Insulated Panels (SIPs) in construction, which offer a range of advantages over traditional building materials. SIPs panels are a light yet durable combination of panel board and either closed-cell polyurethane (PU) or expanded polystyrene (EPS) insulating foam. They are air-tight, and as such provide excellent thermal performance. They also offer superior damp and cold resistance when compared to timber and other materials, and are immune to both compression shrinking and cold bridging. Modular units may also be designed to fit in with external aesthetics of any existing building and modular units once assembled can be virtually indistinguishable from a site-built structure.

- Improved Air Quality - Many of the indoor air quality issues identified in new construction result from high moisture levels in the framing materials. Because the modular structure is substantially completed in a factory-controlled setting using dry materials, the potential for high levels of moisture being trapped in the new construction is eliminated.

Modular buildings can also contribute to LEED requirements in any category site-built construction can, and can even provide an advantage in the areas of Sustainable Sites, Energy and Atmosphere, Materials and Resources, and Indoor Environmental Quality. Modular construction can also provide an advantage in similar categories in the International Green Construction Code.

Disadvantages

- Volumetric: Transporting the completed modular building sections take up a lot of space.

- Flexibility: Due to transport and sometimes manufacturing restrictions, module size can be limited, affecting room sizes.

- Some financial institutions may be hesitant to offer a loan for a modular home.

Market Acceptance

Raines Court is a multi-story modular housing block in Stoke Newington, London,
one of the first two residential buildings in Britain of this type.

Some home buyers and some lending institutions resist consideration of modular homes as equivalent in value to site-built homes. While the homes themselves may be of equivalent quality, entrenched zoning regulations and psychological marketplace factors may create hurdles for buyers or builders of modular homes and should be considered as part of the decision-making process when exploring this type of home as a living and/or investment option. In the UK and Australia, modular homes have become accepted in some regional areas; however, they are not commonly built in major cities. Modular homes are becoming increasingly common in Japanese urban areas, due to improvements in design and quality, speed and compactness of onsite assembly, as well as due to lowering costs and ease of repair after earthquakes. Recent innovations allow modular buildings to be indistinguishable from site-built structures. Surveys have shown that individuals can rarely tell the difference between a modular home and a site-built home.

Modular Homes vs. Mobile Homes

Differences include the building codes that govern the construction, types of material used and how they are appraised by banks for lending purposes. Modular homes are built to either local or state building codes as opposed to manufactured homes, which are also built in a factory but are governed by a federal building code. The codes that govern the construction of modular homes are exactly the same codes that govern the construction of site-constructed homes. In the United States, all modular homes are constructed according to the International Building Code (IBC), IRC, BOCA or the code that has been adopted by the local jurisdiction. In some states, such as California, mobile homes must still be registered yearly, like vehicles or standard trailers, with the Department of Motor Vehicles or other state agency. This is true even if the owners remove the axles and place it on a permanent foundation.

Recognizing a Mobile Or Manufactured Home

A mobile home should have a small metal tag on the outside of each section. If you cannot locate a tag, you should be able to find details about the home in the electrical panel box. This tag should also reveal a manufacturing date. Modular homes do not have metal tags on the outside but will have a dataplate installed inside the home, usually under the kitchen sink or in a closet. The dataplate will provide information such as the manufacturer, third party inspection agency, appliance information, and manufacture date.

Materials

The materials used in modular homes are typically the same as site constructed homes. Wood-frame floors, walls and roof are often utilized. Some modular homes include brick or stone exteriors, granite counters and steeply pitched roofs. Modulars can be designed to sit on a perimeter foundation or basement. In contrast, mobile homes are constructed with a steel chassis that is integral to the integrity of the floor system. Modular buildings can be custom built to a client's specifications. Current designs include multi-story units, multi-family units and entire apartment complexes. The negative stereotype commonly associated with mobile homes has prompted some manufacturers to start using the term "off-site construction."

Financing

Mobile homes often require special lenders. Modular homes on the other hand are financed as site built homes with a construction loan.

Prefabricated Building

Prefabrication, the assembly of buildings or their components at a location other than the building site. The method controls construction costs by economizing on time, wages, and materials. Prefabricated units may include doors, stairs, window walls, wall panels, floor panels, roof trusses, room-sized components, and even entire buildings.

The concept and practice of prefabrication in one form or another has been part of human experience for centuries; the modern sense of prefabrication, however, dates from about 1905. Until the invention of the gasoline-powered truck, prefabricated units—as distinct from precut building materials such as stones and logs—were of ultralight construction. Since World War I the prefabrication of more massive building elements has developed in accordance with the fluctuation of building activity in the United States, the Soviet Union, and western Europe.

Prefabrication requires the cooperation of architects, suppliers, and builders regarding the size of basic modular units. In the American building industry, for example, the 4 × 8-foot panel is a standard unit. Building plans are drafted using 8-foot ceilings, and floor plans are described in multiples of four. Suppliers of prefabricated wall units build wall frames in dimensions of 8 feet high by 4, 8, 16, or 24 feet long. Insulation, plumbing, electrical wiring, ventilation systems, doors, and windows are all constructed to fit within the 4 × 8-foot modular unit.

Another prefabricated unit widely used in light construction is the roof truss, which is manufactured and stockpiled according to angle of pitch and horizontal length in 4-foot increments.

On the scale of institutional and office buildings and works of civil engineering, such as bridges and dams, rigid frameworks of steel with spans up to 120 feet (37 m) are prefabricated. The skins of large buildings are often modular units of porcelainized steel. Stairwells are delivered in prefabricated steel units. Raceways and ducts for electrical wiring, plumbing, and ventilation are built into the metal deck panels used in floors and roofs. The Verrazano-Narrows Bridge in New York City (with a span of 4,260 feet [1,298 m]) is made of 60 prefabricated units weighing 400 tons each.

Precast concrete components include slabs, beams, stairways, modular boxes, and even kitchens and bathrooms complete with precast concrete fixtures.

A prefabricated building component that is mass-produced in an assembly line can be made in a shorter time for lower cost than a similar element fabricated by highly paid skilled labourers at a building site. Many contemporary building components also require specialized equipment for their construction that cannot be economically moved from one building site to another. Savings in material costs and assembly time are facilitated by locating the prefabrication operation at a permanent site. Materials that have become highly specialized, with attendant fluctuations in price and availability, can be stockpiled at prefabrication shops or factories. In addition, the standardization of building components makes it possible for construction to take place where the raw material is least expensive.

Prefabricated Buildings May be of two Types

- Buildings erected directly on the site, where the first set strong skeleton and then delivered insulated sandwich panels and mounted on the prepared skeleton.

- The second method of construction will in time have been factory-created whole modules, up to the necessary communications gasket and the area modules are simply collected in one unit. Thus, the building can be erected in less than a month.

Prefabricated Low-rise Buildings

Especially popular technology rapid construction is to install low-rise buildings, which may be of the following types:

1. Module buildings - consist of pre-engineered unit. Required to deliver a unit on site and install it on a foundation of tape, and then connect the communications. Interior decoration is minimal, since the base of the walls all done at the factory. Container building can be used for temporary accommodation, arrangement of trading pavilions, medical centers, sports facilities and municipal buildings destination. only, but a significant disadvantage of such structures is the limited architectural variation. You can change a number of floors and color units. With regard to life, it is about 50 years.

2. Buildings of the large volume blocks - compared with the "container", they are more expensive, but at the same time create a more comfortable environment for working and living conditions. The blocks are made of a metal frame, followed by finishing sheet materials: sandwich panels, DSP, OSB, particle board, siding and modern insulation. The disadvantage may be called the complexity of the transportation and preservation of intact elements of communications and finishing blocks. The service life of more than half a century, not at the end of which is easily utilized.

3. Large-panel buildings - consist of sandwich panels with excellent thermal and sound insulation qualities. In addition, the panel consisting of sheet metal and insulation are durable, provide an optimal indoor climate and are aesthetically pleasing appearance due to modern finishing coatings. The dimensions depend on the dimensions of the panels erected structure may have a width of 1 meter and a length of 9 meters. The main advantage of prefabricated buildings is the ease of assembly without the use of welding equipment. This technology is also suitable for creating glass facades and three-dimensional translucent structures.

4. Frame buildings - are based on the construction of a metal or wooden frame Weatherproof with skin panels. The construction is carried out directly at the site where the frame is mounted, here it is sheathed and insulated with basalt or mineral heaters. Similarly, it can be erected building of any architectural form and function.

5. Method permanent shuttering – This is another technology used in the construction of low-rise buildings. The method of construction is based on the installation of the formwork, which is poured and the concrete mix. Such houses and buildings are obtained strong and well retain heat. When compared with other technologies, this method takes more time than, for example, block construction.

Prefabricated Steel Buildings

In the construction are increasingly used metal. Metal can withstand heavy loads, has high strength, hardness and long service life. It is not surprising that this material has become an indispensable element in the construction of certain types of prefabricated buildings. In addition to high quality metal has such merits as a low cost, which gives economical construction, compared with wood or other building materials.

The use of metal provided an opportunity to erect a building where it is impossible construction of brick and concrete houses due to their heavy weight. Typically, heavy structures require deep reinforced foundations, which can not be build on plots with the ground does not meet the standards of quality construction. At home, on the basis of the metal frame are lightweight and have enough facilities to melkozaglublennym strip foundation.

Advantages

Well, there are plenty of advantages of using prefabricated buildings.

- Decreased environment-friendly/material waste
- Speed of construction
- Improved quality of materials used
- Simplified on some site logistics

Disadvantages of using Prefabricated Buildings

Like any other construction materials, these buildings are also said to be its share disadvantages for its numerous users. One of disadvantage is that you will certainly not be capable of obtaining a quick loan for such type of construction. This is mainly because not every loan company is searching at these buildings like a traditional structure.

Moreover, if you are fully determined about using these materials for your buildings, then you should take a great look always into your budget. All you need to do is to evaluate the things quite carefully and also make sure that you are generally incorporating the cost of the site work into your budget.

Portable Building

A portable building is a structure that is designed and built to be movable rather than located permanently. A portable building is also called moveable, demountable, transportable and temporary architecture. A mobile home could be classified as a portable building as you can transport the "home" easily from one area to another.

Portable buildings known as *yurts* were used by the nomads of Central Asia centuries ago. Today modern portable homes are designed so that they could be transported by lorry from one place to another and can be lifted and set down by a crane.

The first commercial portable building was introduced in the United Stated in 1955 by Porta-Kamp. In the UK the first portable building was in 1961 under the brand named Portakabin. Today portacabin (with a "c") means portable building in UK.

The population boom after WW II brought about an urgent need for fast and easy housing. This need brought about rise in prefabricated buildings and modular homes. A growth in population meant growth in every aspect. Construction boom soon followed that portable buildings were first conceptualized to quickly provide on-site accommodations for construction workers in remote places. Portable buildings have come a long way. They no longer just provide housing for construction workers but more.

Uses of Portable Buildings

Think of any construction need and chances are any leading portable building provider company can answer your needs. Portable buildings made of steel, designed and built specifically to meet your requirements can be accommodated. There are generic models available for rent or sale but if you want one specifically built to your specifications, it can be done.

Portable buildings have various uses. They could be used as stand-alone construction or as stacked up buildings with metal stairs set outside. These are usually built on construction sites for the construction staff. Portable buildings are also used for offices, classrooms, daycares, doctor's clinic, washrooms or toilet blocks, and more. Portable buildings can also used for storage spaces and garages and even as commercial shops. They could also be used for temporary or even permanent residential units.

Other portable building applications are as kiosks; booths - toll, security, parking; as military camps; and as temporary shelters for disaster victims or the homeless. A portable building is fast and easy to set up. It is very convenient especially if there is an immediate need for its usage. A portable building has all the amenities of a regular house or structure as it is fitted with electrical, sanitary and plumbing services and facilities. Heating and cooling systems are also integrated into each design.

Portable buildings cannot only stand alone but they may be annexed to an existing structure too.

Advantages of using a Portable Building

You might still have memories of ugly steel encasement for a portable building. Though the exterior of a modern day portable building is pretty much the same (steel), different exterior and interior finishing have been applied and integrated into the design to make these buildings aesthetically pleasing. It is hard for the untrained eye to spot which one is a portable building or not especially the stacked ones.

There are several advantages in using portable buildings and they could be divided into three major parts: economic, environmental and social.

Economics of Portable Buildings

- Construction Phase: There is a set pattern of predictability with the construction as the building is factory made, and the is not in any way affected by disruptions caused by weather conditions, unavailability of materials or other site related problems. There is no cost depreciation or appreciation as the portable building is pre-made. Studies show that portable buildings are 99.6% built on time and within budget.

- Construction Speed: A portable building cuts by half the construction time that there is a lot saved on manpower alone. There is also a fast return of investment.

- Quality of Units: Since each and every part is factory made, high quality parts are consistently delivered. There are no irritating and wasteful errors on modules made.

Environment Effects of Portable Buildings

- Building performance is maximized: Factory made modules pass through a rigid quality control inspection. These portable modules are designed to meet the requirements of Building Regulations Part L2 regarding carbon emission. It has been found that most portable buildings are energy efficient thereby reducing energy use.

- Materials used are Ozone-friendly: Materials used for insulations in the roof, walls and floor of most portable buildings have a zero Ozone Depleting Potential. This is good news as using portable building has a minimal impact on the environment.

- Recyclable materials: Portable buildings are obviously recyclable. Bolted parts can easily be dismantled and re-used. As these buildings are made of steel, the major component is highly and even endlessly recyclable.

- Minimized waste: There is less waste generated at the factory and site location. There is also less noise and less vehicle traffic at the site.

- Health and Safety: It is a fact that factory-based construction is safer than on-site construction.

Disadvantages

Portable buildings do not require a foundation, however they may be subject to movement when the ground they are sitting upon freezes and thaws. Smaller buildings may be subject to "rocking" while any portable building could settle, which may lead to doors not closing properly. This problem can be prevented by providing a solid foundation (either footings or a concrete slab) for the building to sit on, while still allowing the building itself to remain portable.

Some areas may require buildings to be tied down to keep them in place in high winds. This is usually a fairly simple task, using auger style earth anchors which screw into the ground, and securing them to your shed. Mobile home tie down kits are also available that will work very well for securing your portable building.

References

- "Factory-Built Construction and the American Homebuyer: Perceptions and Opportunities" (PDF). Huduser. gov. p. 9. Retrieved 2017-09-10

- Modular-building-types: panelbuilt.com, Retrieved 29 April 2018

- Asadi, Pouria (March 2016). "Response modification factor due to ductility of screen-grid ICF wall system in high seismic risk zones". KSCE Journal of Civil Engineering. doi:10.1007/s12205-016-0083-7. ISSN 1226-7988

- Prefabrication, technology: britannica.com, Retrieved 24 May 2018

- Solutions, Dryside Property - Jennifer Mitchell and Magic Web. "Mobile homes vs Manufactured homes vs Modular homes". Drysideproperty.com. Retrieved 2018-03-09

- Prefabricated-buildings-characteristics-and-types, construction-and-infrastructure: helprepair.info, Retrieved 14 July 2018

Allied Aspects of Building Designing

Building design is the process of the application of engineering, architectural and technical principles to the design of buildings. This chapter has been written to provide a comprehensive understanding of the significant aspects of building design such as building insulation, architectural lighting, house wiring, electronic security system, light fixture and carpentry.

Building Insulation

In building construction, various forms of insulation are included as means of reducing the transmission of thermal energy through walls, ceilings, and floors. In simple terms, this helps keep heated interior space warmer in the winter, and air-conditioned spaces cooler in the summer. Reducing the transmission of thermal energy not only makes spaces more comfortable, but it is also essential to controlling energy costs.

All building materials have an inherent insulating property, usually measured on a scale known as R-value, but additional materials are integrated into the building construction specifically to add insulating value to the walls, floors, and ceilings.

Different Types of Insulation: Form

Blanket and Batt Insulation

Glow images/Getty Images

By far the most common type of insulation consists of a "blanket" of rolls or batts of insulation used to fill the cavities between framing members in walls, ceilings, and floors. Batts can also be laid in a blanket across attic floors in to insulate the spaces below.

Batts and rolls consist of flexible fibers woven together in strips of varying widths and thicknesses for different applications. Most commonly, batt insulation uses fiberglass fibers, but it can also be made using mineral fibers, plastic fibers, or natural fibers such as wool or cotton.

Blanket insulation can offer R-values ranging from R-11 (for 3.5-inch-thick batts) to 38 (for 12-inch-thick batts). Batt insulation is one of the most inexpensive forms of insulation, and one of the easiest to install.

Foam Board Insulation

Polystyrene or polyurethane foam boards provide superior insulating R-values for relatively little thickness. They can be used for nearly any part of a building and are especially effective for insulating exterior wall sheathing, interior sheathing for basement walls, and in special applications such as attic hatches or air gaps where floor joists meet foundations. They offer a good way to insulate the spaces between roof rafters in unvented attics.

When used on interior wall applications, foam panels must be covered with half-inch-thick gypsum board panels or another approved building material.

Foam board provides excellent thermal resistance, as much as twice that for other materials of the same thickness. They can easily be trimmed to size for various applications.

Liquid Foam

Liquid foam insulation consists of cementitious or polyurethane materials that are sprayed, injected, or poured into walls or under floors, where it then hardens into an excellent insulating material. It is ideal for irregularly shaped areas and around obstructions, or it can be used to add insulation over existing finished areas. It is a good option for insulating existing walls, as it can be injected into them without removing the wall surfaces.

Liquid foam may allow you to achieve higher R-values than with traditional batt insulation, and it has the advantage of being able to fill the smallest holes to reduce air gaps around pipes, door and window frames, and plumbing and electrical lines.

Spray foam is available in many forms. It can be applied professionally to large areas by contractors using specialized machines or applied to small air gaps using simple spray cans available at home improvement centers.

Loose-Fill Insulation

Loose-fill and blown-in insulation, usually consisting of cellulose, fiberglass, or mineral wool, can be blown or poured into the joist cavities on attic floors or in the joist cavities of walls.

R-values vary depending on material, and loose fill insulation has a tendency to settle over time, reducing its R-value. But it is relatively inexpensive and is considered a "green" option since these materials are created from recycled waste materials. Cellulose insulation is made mostly from recycled newsprint, most fiberglass insulation is made from 40 to 60 percent recycled glass, and mineral wool contains about 75 percent recycled material.

Concrete Block Insulation

Concrete and concrete block walls can be insulated in a number of ways.

Concrete block can be insulated by applying rigid foam board to either the outside of the walls (on new construction) or the interior walls (on existing homes).

An additional method of insulating concrete blocks involves the use of blocks consisting of autoclaved aerated concrete (AAC) or autoclaved cellular concrete (ACC). These materials contain about 80 percent air by volume and have about 10 times the insulating value of traditional concrete blocks. Precast ACC blocks use fly ash instead of the high-silica sand used in AAC blocks. Autoclaved blocks easily absorb moisture, so they must be protected from water, but they are lightweight and easy to install.

In poured concrete foundations, beads of polystyrene foam can also be incorporated into the concrete mix to increase their R-value. This method can increase the R-value 10-fold over standard poured concrete.

Radiant Barriers and Reflective Insulation

While most insulations work by resisting conductive and convective heat flow, reflective insulation works by actually reflecting back radiant heat.

These insulations incorporate a radiant material—usually shiny aluminum foil—applied to a layer of traditional insulation that also has some form of backing—kraft paper, plastic film, or cardboard. It is most commonly used in attics to reduce summer heat gain and to lower cooling costs. It is one of the best types of insulation for preventing downward heat flow.

Reflective insulation forms a radiant barrier that lessens heat transfer from roofs down into an attic space. It must face an air space in order to be effective. They are most effective in hot climates, where they can lower cooling costs by 5 to 10 percent. However, in cooler climates, traditional thermal insulation is a better choice.

Rigid Fiberboard Insulation

Rigid fiberboard made from either fiberglass or mineral wool is normally used in places that will be affected by high temperatures such as the ductwork for HVAC systems. One of the advantages of this type of insulation is that it can be preinstalled on ductwork at shops or custom-fabricated at the job sites. The panels come in a range of thicknesses, from 1 to 2.5 inches. Both faced and unfaced boards are available.

Fiberboard insulation is usually installed by HVAC contractors, who apply it to exterior duct surfaces with a system of pins or clips. When unfaced boards are used, the outer surfaces are finished with an insulating cement or canvas. With faced boards, the joints between panels are sealed with tape or mastic.

Structural Insulated Panels (SIPs)

An entirely different method of insulating a building can be achieved with the use of structural insulated panels (SIPs) instead of traditional stud framing. SIPs are large prefabricated panels that include a 4- to 8-inch thick foam board insulating material (usually polystyrene or polyisocyanurate) sandwiched between strong facing materials, such as oriented strand board (OSB). An SIP-constructed building can be as much as 12 to 14 percent more energy efficient than a traditional "stick-frame" home. An SIP building will also be more airtight and quieter.

SIPs are not an option for insulating existing buildings, but can be considered when planning a new building or major addition. In addition to offering excellent insulating value, SIPs are structurally stronger and more stable than traditional framing.

Insulating Concrete Forms (ICFs)

Insulating concrete forms (ICFs) are prefabricated forms for poured concrete walls that remain as part of the wall assembly. The system consists of foam boards or interlocking foam insulation blocks joined together with plastic ties. When concrete is poured into the foams, the resulting walls achieve an insulating value of about R-20.

ICFs can be used for foundation construction alone in buildings with basements, or they can form the above-ground walls as well. Structures built from ICFs still resemble traditional framed structures.

Installing an ICF system is a highly specialized skill that requires an experienced contractor. Especially critical is making sure the walls are insect- and waterproof.

Different Types of Insulation: Materials

Fiberglass

Fiberglass Insulation.

Fiberglass is the most common insulation used in modern times. Because of how it is made, by effectively weaving fine strands of glass into an insulation material, fiberglass is able to minimize heat transfer. The main downside of fiberglass is the danger of handling it. Since fiberglass is made out of finely woven silicon, glass powder and tiny shards of glass are formed. These can cause damage to the eyes, lungs, and even skin if the proper safety equipment isn't worn. Nevertheless, when the proper safety equipment is used, fiberglass installation can be performed without incident.

Fiberglass is an excellent non-flammable insulation material, with R-values ranging from R-2.9 to R-3.8 per inch. If you are seeking a cheap insulation this is definitely the way to go, though installing it requires safety precautions. Be sure to use eye protection, masks, and gloves when handling this product.

Mineral Wool

Mineral Wool.

Mineral wool actually refers to several different types of insulation. First, it may refer to glass wool which is fiberglass manufactured from recycled glass. Second, it may refer to rock wool which is a type of insulation made from basalt. Finally, it may refer to slag wool which is produced from the slag from steel mills. The majority of mineral wool in the United States is actually slag wool.

Mineral wool can be purchased in batts or as a loose material. Most mineral wool does not have additives to make it fire resistant, making it poor for use in situation where extreme heat is present. However, it is not combustable. When used in conjunction with other, more fire resistant forms of insulation, mineral wool can definitely be an effective way of insulating large areas. Mineral wool has an R-value ranging from R-2.8 to R-3.5.

Cellulose

Cellulose insulation is perhaps one of the most eco-friendly forms of insulation. Cellulose is made from recycled cardboard, paper, and other similar materials and comes in loose form. Cellulose has an R-value between R-3.1 and R-3.7. Some recent studies on cellulose have shown that it might be an excellent product for use in minimizing fire damage. Because of the compactness of the material, cellulose contains next to no oxygen within it. Without oxygen within the material, this helps to minimize the amount of damage that a fire can cause.

So not only is cellulose perhaps one of the most eco-friendly forms of insulation, but it is also one of the most fire resistant forms of insulation. However, there are certain downsides to this material as well, such as the allergies that some people may have to newspaper dust. Also, finding individuals skilled in using this type of insulation is relatively hard compared to, say, fiberglass. Still, cellulose is a cheap and effective means of insulating.

Cellulose Insulation Material.

Polyurethane Foam

Polyurethane Insulation.

While not the most abundant of insulations, polyurethane foams are an excellent form of insulation. Nowadays, polyurethane foams use non-chlorofluorocarbon (CFC) gas for use as a blowing agent. This helps to decrease the amount of damage to the ozone layer. They are relatively light, weighing approximately two pounds per cubic foot (2 lb/ft^3). They have an R-value of approximately R-6.3 per inch of thickness. There are also low density foams that can be sprayed into areas that have no insulation. These types of polyurethane insulation tend to have approximately R-3.6 rating per inch of thickness. Another advantage of this type of insulation is that it is fire resistant.

Polystyrene

Polystyrene (Styrofoam).

Polystyrene is a waterproof thermoplastic foam which is an excellent sound and temperature insulation material. It comes in two types, expanded (EPS) and extruded (XEPS) also known as Styrofoam. The two types differ in performance ratings and cost. The more costly XEPS has a R-value of R-5.5 while EPS is R-4. Polystyrene insulation has a uniquely smooth surface which no other type of insulation possesses.

Typically the foam is created or cut into blocks, ideal for wall insulation. The foam is flammable and needs to be coated in a fireproofing chemical called Hexabromocyclododecane (HBCD). HBCD has been brought under fire recently for health and environmental risks associated with its use.

Other Common Insulation Materials

Although the items listed above are the most common insulation materials, they are not the only ones used. Recently, materials like aerogel (used by NASA for the construction of heat resistant tiles, capable of withstanding heat up to approximately 2000 degrees Fahrenheit with little or no heat transfer), have become affordable and available. One in particular is Pyrogel XT. Pyrogel is one of the most efficient industrial insulations in the world. Its required thicknesses are 50% – 80% less than other insulation materials. Although a little more expensive than some of the other insulation materials, Pyrogel is being used more and more for specific applications.

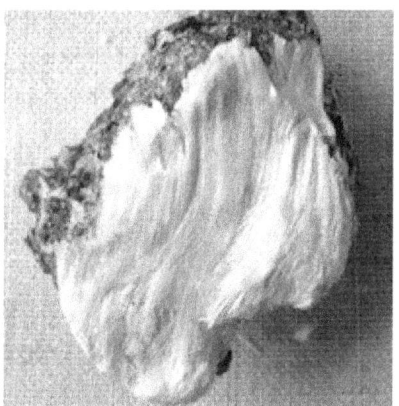

Asbestos.

Other insulation materials not mentioned are natural fibers such as hemp, sheep's wool, cotton, and straw. Polyisocyanurate, similar to polyurethane, is a closed cell thermoset plastic with a high R-value making it a popular choice as an insulator as well. Some health hazardous materials that were used in the past as insulation and are now outlawed, unavailable, or uncommonly used are vermiculite, perlite, and urea-formaldehyde. These materials have reputations for containing formaldehyde or asbestos, which has essentially removed them from the list of commonly used insulation materials.

Different Types of Insulation: Locations

Attic Insulation

If you fail to install attic or loft insulation, your home—particularly the second story—will be much hotter and more uncomfortable in the summer. Insulation or reinsulating your attic helps reduce your utility bills, give your home more consistent temperatures throughout, reduce noises on different levels inside, and contribute to a healthier indoor air quality inside your home.

Wall Insulation

When it comes to home insulation, your walls are some of the best places to insulate. When it comes to insulating walls, there are a couple things to keep in mind: Exterior wall insulation *and* interior wall insulation are essential to maximizing your home's energy efficiency and muffling noises from both outside and inside, between different rooms and levels.

Crawl Space Insulation

For such a small area, a crawl space can have a big huge impact on your home's energy efficiency if not properly insulated. No matter what type of crawl space you have and what material you use, you'll notice the difference crawl space insulation can make.

Floor Insulation

Subflooring and floors over garages should be insulated. Heat enters and escapes buildings through the flooring, so installing insulation is extremely important. Make sure your floors are well insulated so you can help maximize energy savings and year-round comfort.

Ceiling Insulation

Whether you want to add additional soundproofing and insulation between levels or you need to add a barrier between the attic and the rest of the house, ceiling insulation is a smart choice.

Architectural Lighting

Architectural lighting" can be seen as the type of lighting that arises from and for architecture. In a sense, architectural lighting is directly opposite decorative lighting. This contradiction can mainly be deduced from the thinking that precedes the preparation of a lighting plan.

Lighting plays a vital role in the way people experience and understand architecture. Whether buildings and structures are lit naturally or artificially, lighting is the medium that allows us to see and appreciate the beauty in the buildings around us.

Lighting can bring an emotional value to architecture – it helps create an experience for those who occupy the space. Without lighting, where would architecture be? Would it still have the same impact? No, it wouldn't. Whether it's daylighting or artificial lighting, light draws attention to textures, colors, and forms of a space, helping architecture achieve its true purpose. Vision is the single most important sense through which we enjoy architecture, and lighting enhances the way we perceive architecture even more.

To create a successful balance between lighting and architecture, it's important to remember three key aspects of architectural lighting: (1) aesthetic, (2) function, and (3) efficiency.

1. Aesthetics – Lighting schemes can influence the aesthetic enjoyment and sensory appreciation of many different types of spaces, including:

Retail – Carefully placed lighting systems can highlight aspirational items and invite customers to linger, browse and buy.

Public – Parks, town centres and entertainment districts can use lighting to create a welcoming and vibrant environment.

Private – Office buildings can use creative lighting concepts in various imaginative ways to represent their brand image.

2. Energy Efficiency – To ensure light is not wasted designers must strive to install lighting systems that have maximum impact with minimum energy consumption. Energy efficient lights can be directed at key features or at reflective, shiny objects to make the most of the light. Clever design will minimise the amount of lighting required.

3. Functionality – Ergonomics of lighting fixtures is the final key consideration. Lighting needs to produce sufficient light for the purposes of the structure while balancing factors like operating costs and how that light will be perceived and interacted with by people.

House Wiring

House wiring is defined as any wiring or electrical system used in a home or its surrounding areas. The wiring process is fairly time consuming and requires planning for the varying power needs of electronics and appliances. In a home, the wiring system includes outlets, the main panel and meter base, and it is essential that all pieces are installed and function together properly to keep the home safe. With this in mind, hiring a professional is usually best to ensure that the process is completed safely and to code.

Wiring Diagrams

A wiring diagram is an essential tool for beginning electricians. This guide outlines the wiring process and provides information about control panels and where all elements of the electrical system will be placed. These diagrams are available from a variety of online sites and from do-it-yourself wiring books. A good guide will include illustrations with information on wire color and specifics on where each wire should end up upon completion of the project. Guides that are printed from an online source should have type that is legible and easy to read quickly in case a problem arises. This guide should be kept nearby throughout the project.

Service Equipment

It is important to become familiar with local building requirements, particularly when it comes to electrical service equipment. This includes placement of the main panel, meter base and conductors, each of which must be wired with the appropriate voltage for the space. More voltage is often required for the kitchen, as it is home to several heavy-duty appliances. It is essential that main panels be supplied with a minimum clearance that is 30 inches wide and 36 inches deep, which leaves room for repairs if a problem occurs. The National Electrical Code also dictates that enclosed spaces, including closets and bathrooms, are not appropriate spots for main panels.

Outlets

Proper outlet spacing is an essential part of proper wiring and function of the home. Local building codes dictate the spacing between outlets, as well as the type of outlet required for different appliances. Certain appliances, particularly the dishwasher and refrigerator, require more voltage than other parts of the home and should be run with larger gauge wire. In the kitchen, all countertops and any area designated for eating must be equipped with a style of outlet designed to prevent shocks and electrocution.

Elements

Power Point

Power points (receptacles, plugs, wallsockets) need to be installed throughout the house in locations where power will be required. In many areas the installation must be done in compliance with standards and by a licensed or qualified electrician. Power points are typically located where there will be an appliance installed such as, telephone, computers, television, home theater, security system, CCTV system.

Light Fittings and Switches

The number of light fitting does depend on the type of light fitting and the lighting requirements in each room. The incandescent bulb made household lighting practical, but modern homes use a wide variety of light sources to provide desired light levels with higher energy efficiency than incandescent lamps. A lighting designer can provide specific recommendations for lighting in a home. Layout of lighting in the home must consider control of lighting since this affects the wiring. For example, multiway switching is useful for corridors and stairwells so that a light can be turned on and off from two locations. Outdoor yard lighting, and lighting for outbuildings such as garages may use switches inside the home.

Telephone

Telephone wiring is required between the telephone company's service entrance and locations throughout the home. Often a home will have telephone outlets in the kitchen, study, living room or bedrooms for convenience. Telephone company regulations may limit the total number of telephones that can be in use at one time. The telephone cabling typically uses two pair twisted cable terminated onto a telephone plug. The cabling is typically installed as a daisy chain starting from the point where the telephone company connects to the home or outlets may each be wired back to the entrance.

Data

Data wiring has two components, these are:

1. Data service delivery

2. Data network cable

The three most common ways data services are delivered to the home:

1. ADSL service on the back of the telephone cabling

2. Cable Modem

3. Fiber

ADSL Service

ADSL services are typically delivered using the telephone cabling. An ADSL modem needs a filter to segregate voice handsets from the ADSL modem.

Cable Modem cable modems are typically installed in location where there is an existing Pay TV service outlet. The installation requires the installation of a Pay TV outlet (F connector).

Fiber Fiber is the least common but it is growing in numbers. If the home has fiber to it then the fiber terminates on what is known as an Optical Network Termination unit (ONT) and it has a data port on it. Cabling from the street to the point where the ONT is installed is fiber and is typically installed by the service provider.

In all three cases the equipment supplied by the Internet provider will have a connection to the computers installed in the building. This is the data network cabling or LAN cabling.

If more than one computer or device (PC, printers, TV etc.) is to be connected in the home, LAN cabling will be required. The cabling used for data networking is similar to the phone cabling as it is twisted pair but of a much higher quality. The cable is known as Category (Cat) 5 or Cat 6. The cabling must be installed as a star wired configuration, that is the cabling runs from the point next to the modem, hub, or router uninterrupted up to the outlet next to the device that needs to be connected. Computer network wiring cannot be chained from one outlet to the next; each outlet is wired individually back to the hub or router next to the modem. If only one computer is required,it can be directly plugged into the modem. An alternative to a wired LAN especially useful for mobile devices is a wireless LAN, which can reduce or eliminate all the fixed wiring.

Television

Cabling for free to air TV requires the following:

- An antenna
- Coaxial cable
- TV outlets

Antenna types vary depending on location; an urban area with nearby transmitters will require a smaller antenna than a rural site with distant stations. The antenna is often mounted outdoors on the roof or a tower. A coaxial or twinlead cable is run from the antenna to the location where the television is located. One common type of cable is designated RG-6 Tri-shield or quad-shield cable. The cable is terminated on a television outlets, typically an F connector mounted on a face plate. If there are multiple outlets, an RF splitter is used to divide the signal among them; outlets on the splitter are connected to television outlets at each location (living room, rec room, bedrooms, den, for example). RF splitters come with different types; some include amplifiers for multiple outlets.

Whilst most TV outlets use the F connector the Television or digital set top box usually come with a connector known as Belling Lee so the cable used to connect from the TV outlet to the television will need to have an F connector in one end and a Belling Lee connector at the other end.

The distribution of pay TV through the home uses the same type of cabling used for Free to Air TV with some variations. The variations are:

1. There is no antenna as there is either a satellite dish or a cable from the street.
2. The cabling must be RG-6 quad shield.
3. You may be required to use the cable and cabling connectors approved by your pay TV provider
4. A Pay TV Set Top Box needs to be installed at each television where you want to have access to Pay TV services.

In most cases the pay TV company will supply and install the satellite dish or cable from the street and the cabling to the TV set. In many cases Pay TV services also require a telephone point to access movies on demand.

IPTV is television delivered to the home over the Internet. Any device for viewing IPTV must have an internet connection. This may be a wired connection, or wireless.

Home Theater

Home theater pre-wiring requires knowledge of the number of speakers to be installed.

- Two front speakers; one on the left of the screen and one on the right of the screen,
- One front speaker cable just above or below the screen which is the middle front,

- Two rear speakers; one on the left and one of the right in line with front left and right speaker locations,

- The sub-woofer which can be anywhere in the room acoustically but must be relatively close to the active equipment the amplifier or surround sound receiver.

Speaker cable is figure eight multi-strand copper cable. Cabling for the sub-woofer is typically a single shielded cable terminated on an RCA connector. A 7.1 channel system also needs cable for speakers that are installed between the front and back speakers.

The simplest layout for a home theater system is a single piece of furniture containing all one's AV equipment, which simplifies wiring. If, on the other hand, a front projection unit is to be employed, more thought must be given to the layout of the system. Several different cabling systems are commonly used for this application, including HDMI, DVI, and VGA.

Distributed Audio

Distributed audio provides music throughout the house, where the music sources are all centralized. Rooms are provided with speakers and controls to adjust volume or music source. A system may have central controls or allow for off-site control.

Security Monitoring

Security monitoring (burglar alarm) systems contain basic components of:

- Code pad

- Siren and strobe light

- Motion detectors

- Main panel

and may have additional components.

Cabling for Traditional Equipment

Code pad The code pad is typically found inside the front door or any other access door. The code pad is used to alarm the system on departure and disarm the system on entry. The cabling required is 6 core multi strand copper cable.

Siren and strobe light The siren and strobe light are typically installed outside the front of the house where it can be seen from the street and is protected from the weather. The cabling required is a 6 core multi strand copper cable.

Motion detectors The motion detectors installed in locations throughout the house were any intrusion into the home can be detected. The best way to think of this is, which are the rooms that have direct access from the outside, where can I place a detector to pick up any intrusion. One solution is to place a motion sensor in each room, as this can be expensive an alternative is place one immediately outside in the common corridor to all rooms. The cabling required is a 6 core multi strand copper cable.

Main equipment. The main equipment is typically installed in a location that is not easily accessible such as a cupboard or sub floor area where in the event of an intrusion the person(s) cannot easily find it and interfere with the unit. The main unit requires a power point installed next to it for main power. It also needs a connection to the telephone line servicing the home so in situations where a back to base service is required it can be connected to the phone line. Note the connection of the security system to the phone line requires a wiring configuration that allows the security system to disconnect all phones in the home when it needs to connect to the monitoring center. This is critical, if the wiring is not correct the system may not communicate back to base when an intrusion is detected.

All cabling from the code pad, siren and strobe light and motion detectors need to be run out from the main equipment. It is also recommended that the cabling to each code pad, motion detector are individual runs from the main equipment to the device. By having each device individually connected to the main equipment is facilitates maintenance and allows for more effective monitoring.

Cabling for IP based Systems

Like the traditional equipment the IP based systems require as a minimum:

1. Code pad

2. Siren and strobe light

3. Motion detectors

4. Main equipment

The difference here is the cabling to connect the main equipment is either Cat 5 or Cat 6 and it is installed as part of the data cabling of the home.

Security CCTV

This is becoming more sought after in private home as an additional level of security. The wiring required to install a CCTV system is Data cabling "Data network cabling". What you need to determine is where do you want to install the CCTV cameras and wherever you want the camera you need to install a data outlet. The location where you install the cameras will vary from home to home but typically they are installed so you can see anyone approaching any of the entry areas of the home.

The advantage of an IP bases system is the flexibility to add devices at a later stage. That is you can cable to as many locations as you want and have it terminate on a data outlet near where you may be planning to add devices at a later stage. Adding the device is as simple as plugging it into the outlet and configuring the device.

Automation

Automation refers to the ability to be able to control a range of devices in the home ranging from lights to curtains. The most common example of automation are referred to as Lighting control systems. Lighting control system need to be installed by a qualified professional as the cabling is only one element but without the equipment and programming you cannot even turn a light on. The cabling required when installing an automation system can be divided into two parts:

1. Electrical

2. Data Bus

Electrical This is cabling installed from the electrical switchboard to the light fitting or any other device that is to be controlled by the automation system. For example, if you have four down lights in a room and you wish to control each light individually, then each light will be wired back using electrical cabling back to the electrical switchboard. This means you will have four electrical cables installed from the electrical switchboard to the location where the light fittings will be installed. Each cable will be a three core active, neutral and earth cable. If in that room you also have a free standing lamp plugged into a power point and you also want to control this from your automation system, you will need to have that power point individually wired back to the electrical switchboard. So if you want to individually control every light fitting and every power point or power outlets then each one of these devices must be individually wired back to the electrical switchboard. As you can see this start to become quite a lot of electrical cabling so planning is essential.

Note, when you are using an automation system, there is no need to install any electrical cabling to the light switches. In a traditional electrical installation without automation the lights in a room would be wired back to the light switch which in turn would be wired back to the switchboard or some similar arrangement, so keep reading.

Data Bus Once you have installed the electrical cabling you need to install the data bus cable from the electrical switchboard to every location you want to have a light switch or control panel installed (control panel is like the code pad on a security system or touch screen that gives you access to various control functions). The most common cable used for this is a Category 5 cable. The cable can be installed in either a daisy chain or star wired configuration. The importance is to minimize the cable length to avoid an communications problem on the bus.

Energy Management

Energy management is a new and upcoming topic in particular at the home. Older systems tended to be cable however all new systems use one of a variety of wireless solutions. This enables them to be effectively retrofitted into existing homes with the minimum of disruption.

If a cabled system is selected cabling needs to be deployed to the major appliances in the home. The cabling is installed as part of the data cabling "Data Network Cabling". In addition to a cable being installed to every major appliance you also need to install a data cable near the electricity meter.

The major appliances being considered at this stage are:

1. Electric hot water system

2. Air Conditioning

3. Pool pump

4. Fridge / freezer

5. Electric vehicle charger

6. Battery energy storage systems (BESS)

Should a wireless system be selected the need for such disruption is removed. Smart plugs or switches can be used to connected the major appliances to the electricity supply and the home energy management system will wirelessly control them.

Wiring Basics

Wiring a residence requires the use of three different colors of wire in addition to a bare ground wire. The bare wire serves as the ground and doesn't transfer any power. This is the wire that will trip the circuit breaker in the event of a short circuit and eliminate any electrical current moving to the device. Both black and red wires are called "hot" wires in the industry and carry current from the breaker to the appliance. White wires hold current taking the opposite path, from the appliance to the breaker and are called "returns." A 14-2 gauge wire is the minimum in most areas, but 12-2 is preferable. This is a thicker gauge wire that will handle more power, which means reduced chance of tripping a circuit. This additional power is an ideal choice for homes that have many appliances or electrical devices that will that run simultaneously.

Wiring Terminology

It helps to understand a few basic terms used to describe wiring. An electrical wire is a type of conductor, a material that conducts electricity. In the case of household wiring, the conductor itself is usually copper or aluminum, and either solid or stranded wire. Most wires in a home are insulated, meaning they are wrapped in a nonconductive plastic coating. One notable exception is ground wires, which are typically solid copper and are either insulated with green insulation or are uninsulated (bare).

The most common type of wiring in modern homes is in the form of nonmetallic (NM) cable, which consists of two or more individual wires wrapped inside a protective plastic sheathing. NM cable usually contains one or more "hot" (current-carrying) wires, a neutral wire, and a ground wire.

As an alternative to NM cable, individual wires can be installed inside of a rigid or flexible metal or plastic tubing called conduit. Conduit is typically used where wiring will be exposed and not hidden inside walls, floors, or ceilings.

NM Cable

Commonly called "Romex," after the popular brand name, NM cable is designed for interior use in dry locations. Almost all of the wiring in a modern home is NM cable. The most common sizes and their amperage (amp) ratings are:

- 14-gauge (15-amp circuits)

- 12-gauge (20-amp circuits)

- 10-gauge (30-amp circuits)

- 8-gauge (40-amp circuits)

- 6-gauge (55-amp circuits)

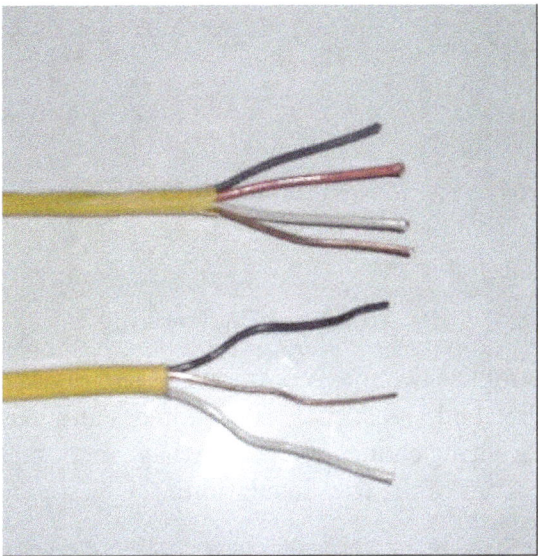

Two types of NM cable.

UF Cable

Underground Feeder (UF) is a type of nonmetallic cable designed for wet locations and direct burial in the ground. It is commonly used for supplying outdoor fixtures, such as lampposts. Like standard NM cable, UF contains insulated hot and neutral wires, plus a bare ground wire. But while sheathing on NM cable is a separate plastic wrap, UF cable sheathing is solid plastic that completely surrounds each wire.

Thhn/Thwn Wire

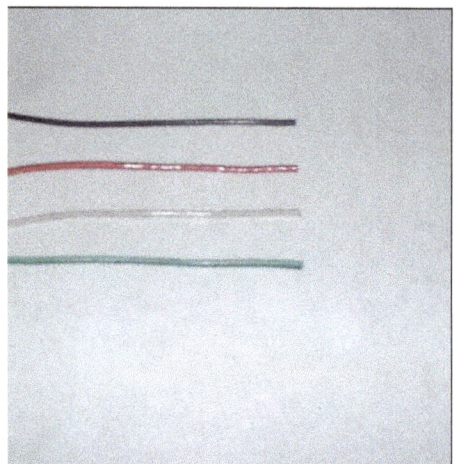

THHN wire. Tim Thiele

THHN and THWN are codes for the two most common types of insulated wire used inside conduit. Conduit is often used in unfinished areas, such as basements and garages, and for short exposed runs inside the home, such as wiring connections for garbage disposers and hot water heaters. The letters indicate specific properties of the wire insulation:

- T: thermoplastic

- H: heat-resistant; HH means highly heat-resistant

- W: rated for wet locations

- N: nylon-coated, for added protection

Low-Voltage Wire

Home Depot

Low-voltage wiring is used for circuits typically requiring 50 volts or less. Three common types are landscape lighting wire, bell wire (for doorbells), and thermostat wire. Wire sizes range from about 22 gauge to 12 gauge. Low-voltage wires typically are insulated and may be contained in cable sheathing or combined in pairs, similar to lamp cord wire. It must be used only for low-voltage applications.

Phone and Data Wire

Telephone and data wiring are low-voltage wires used for "land line" telephones and for internet hookups. Telephone cable may contain four or eight wires. Category 5, or Cat 5, cable, the most common type of household data wiring, contains eight wires wrapped together in four pairs. It can be used for both phone and data transmission and offers greater capacity and quality than standard phone wire.

Electronic Security System

Electronic security system refers to any electronic equipment that could perform security operations like surveillance, access control, alarming or an intrusion control to a facility or an area which uses a power from mains and also a power backup like battery etc. It also includes some of the operations such as electrical, mechanical gear. Determination of a type of security system is purely based on area to be protected and its threats.

Role of Electronic Security System

Electronic security relates to leveraging innovation in defensive holding by anticipating unapproved access to individuals and property. The government is a universal and major customer of such security administrations and business sections also utilizes the security systems for their workers for giving security. These days, one can witness their usage in range like domestic application and small stores moreover.

The electronic security systems extensively comprises of alarms, access controls and CCTVs (close-circuit televisions), which are prominently and broadly utilized. CCTVs have picked up additional significance from all of these products.

Importance of Electronic Security System

The electronic security systems are broadly utilized within corporate work places, commercial places, shopping centers and etc. These systems are also used in railway stations, public places and etc. The systems have profoundly welcomed, since it might be worked from a remote zone. And these systems are also utilized as access control systems, fire recognition and avoidance systems and attendance record systems. As we are know that the crime rates are increasing day by day so most of the people are usually not feeling comfort until they provide a sure for their security either it may be at office or home. So we should choose a better electronic system for securing purpose.

Classification of Electronic Security System

Classification of security system can be done in different ways based on functioning and technology usage, conditions of necessity accordingly. Based on functioning categorizing electronic security system as follows:

- CCTV Surveillance Security System
- Fire Detection/Alarming System
- Access Control/Attendance System

CCTV Surveillance Systems

It is the process of watching over a facility which is under suspicion or area to be secured; main part of the surveillance electronic security system consists of camera or CCTV cameras which forms as eyes to surveillance system. System consists of different kinds of equipment which helps in visualizing and saving of recorded surveillance data. The close-circuits IP cameras and CCTVS transfers image information to a remote access place. The main feature of this system is that, it can use any place where we watch the human being actions. Some of the CCTV surveillance systems are cameras, network equipments, IP cameras and monitors. In this system, we can detect the crime through the camera, the alarm rings after receiving the signal from the cameras which are connected CCTV system; to concern on the detection of interruption or suspicion occurrence on a protected area or capability, the complete operation is based on the CCTV surveillance system through internet. The figure below is representing the CCTV Surveillance Systems.

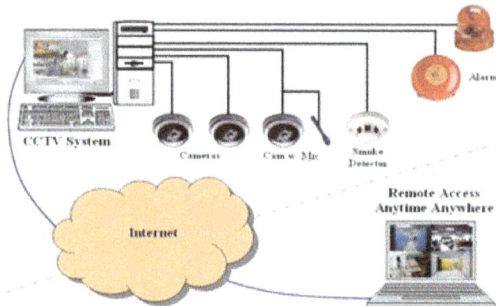

CCTV Surveillance System

- IP Surveillance System

The IP-Surveillance system is designed for security purpose, which gives clients capability to control and record video/audio using an IP PC system/network, for instance, a LAN or the internet. In a simple way, the IP-Surveillance system includes the utilization of a system Polaroid system switch, a computer for review, supervising and saving video/audio, which shown in figure below.

In an IP-Surveillance system, a digitized video/audio streams might be sent to any area even as far and wide as possible if wanted by means of a wired or remote IP systeWm, empowering video controlling and recording from anyplace with system/network access.

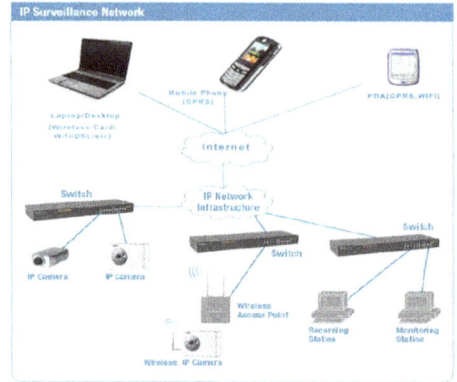

IP Surveillance Network

Fire Detection and Alarming Systems

It can also referred as a detection and alarming system as it gives an alarming alert to concern on detection of interruption or suspicion happening on a protected area or facility. System generally consists of detector using a sensor followed by an alarm or an alerting circuit. The main function of this system is to rapidly extinguish an advancing fire and alarm tenants preceding impressive harm happens by filling the secured zone with a gas or concoction smothering operator. Different types of sensors are available for detection but usage of sensor is purely based on application requirement, like home automation, ware house fire detection, intrusion alert etc.

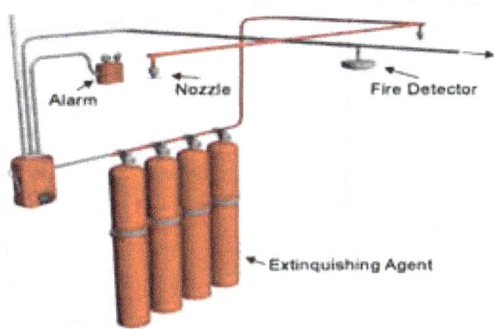

Fire Detection and Alarming system

Attendance and Access Control Systems

System which provides a secured access to a facility or another system to enter or control it can be called as an access control system. It can also act as attendance providing system which can play a dual role. According to user credentials and possessions access control system is classified, what a user uses for access makes system different, user can provide different types like pin credentials, biometrics or smart card. System can even use all possessions from user for a multiple access controls involved. Some of the attendance and access control systems are:

- Access Control System

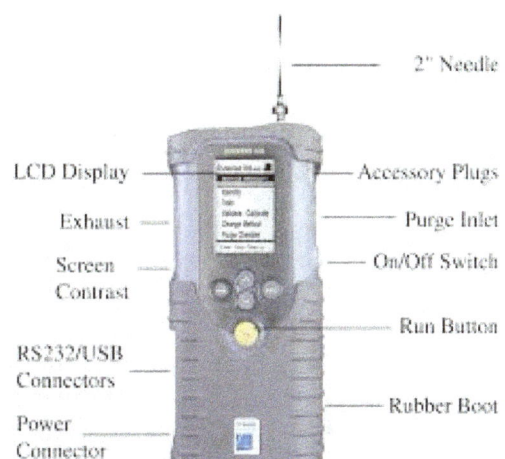

- RF based access control and attendance system

Finger Print Attendance-Access Control System

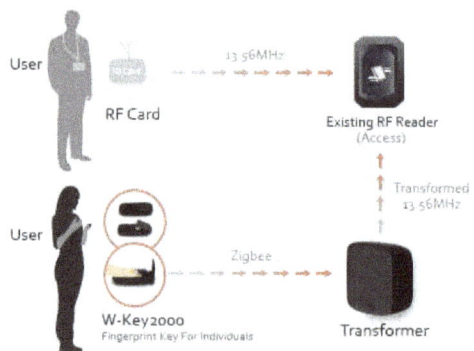

RF based Card access control and attendance system

Applications of Electronic Security System

Electronic security system extends its applications in various fields like home automation, Residential (homes and apartments), commercial (offices, banks lockers), industrial, medical, and transportations. Some of the applications using electronic security system are electronic security system for railway compartment, electronic eye with security, electronic voting system are the most commonly used electronic security system.

One of The Examples Related to Electronic Security System

From the block diagram, the system is mainly designed based on Electronic eye (LDR sensor); we use this kind of systems in bank lockers, jewelry shops. When the cash box is closed, the neither buzzer nor the binary counter/divider indicates that box is closed. If anyone tries to open the locker door then automatically a light falls on the LDR sensor then the resistance decreases slowly this cause buzzer to alert the customer. This process continues until the box is closed.

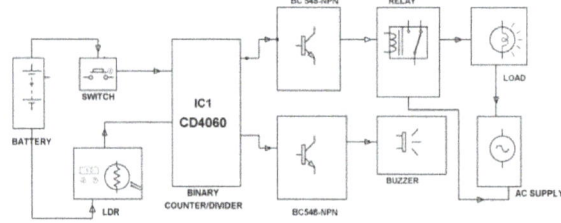

Electronic Eye Controlled Security System

Light Fixture

Light fixtures, also known as luminaires, come in a variety of sizes, shapes, wattage, etc. The diagram below shows the major types of fixtures. Obviously, there are many varieties of each type of fixture.

Descriptions of the types of fixtures are listed below the diagram.

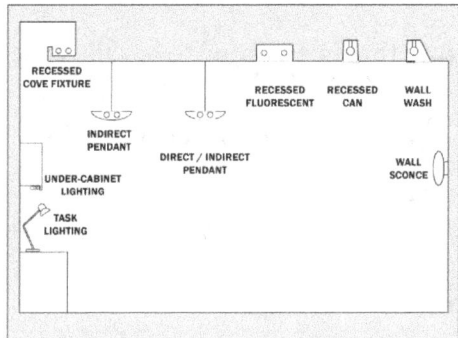

Recessed Cove Lighting Fixture

A recessed cove fixture is mounted in a light cove, which is built above the ceiling at the intersection of the ceiling and the wall. These fixtures typically direct the light toward the wall. It is important that trim at the edge of the cove is tall enough to hide the lighting fixture otherwise the lamp may be visible, which is unappealing.

Indirect Pendant Lighting Fixture

Indirect pendants hang from the ceiling and are usually suspended from cables. The lamp is completely hidden from below and a reflector directs all of the light up toward the ceiling. This type of fixture offers a softer and more even distribution of light within the space. It is best used for general lighting and is not appropriate for task lighting. The ceiling color should be light so that it reflects as much light as possible.

Direct/Indirect Pendant Lighting Fixture

A direct/indirect pendant also hangs from the ceiling, but it directs light up and down. These types of fixtures are used a lot in offices where general lighting is required, but there is also a need for task lighting immediately below the fixture. Direct / indirect fixtures are designed in variations that distribute differing amounts of light up and down so that a designer has control over the light distribution.

Recessed Lighting Fixtures

Recessed lighting fixtures are the most common fixtures used in commercial and institutional construction. Fixtures are sized to work with common ceiling tile sizes. While fluorescent lamps have been most popular in the past, many facilities are shifting to LED fixtures because they last longer and require less maintenance.

Wall Wash Light Fixtures

Wall wash fixtures are recessed lights with reflectors that direct the light toward the wall. They are most often used to highlight art, signage, or other items on a wall.

Wall Sconce

A wall sconce is a decorative fixture that is mounted to a wall. They provide general room lighting, but are mostly decorative.

Task Lighting

Task lighting is a generic description for lights that are used to illuminate specific tasks or work that is being done. Task lights can be lamps, lights mounted to desks, under cabinet lights, or any lighting that helps people see their work better.

Under Cabinet Light Fixtures

Lights are often mounted below cabinets so that tasks on the counter below can be seen easily. These can be found in homes and offices and are generally controlled from a nearby switch or a switch on the light fixture.

Carpentry

The homes we live in and the furniture we sit on would not have been made possible without the skilled trade known as carpentry. The field involves cutting, shaping and putting together materials to be able to build houses, offices, ships and furniture, among others. Wood or timber is the main material used in carpentry, although there are many other materials that are now used in construction as well.

Carpentry can be categorized in two types: 1) Rough Carpentry and 2) Finish Carpentry. Rough carpentry refers to the type of carpentry that does not require a fine finish. These include making the structural parts of the house like the posts, rafters and beams. Rough carpentry also involves making the roofing and framing of a structure. This type of carpentry does not need to be finely finished because the parts are going to be covered anyway.

Finish carpentry, on the other hand, refers to the flooring, staircases, window installation and moldings and trims on buildings, houses and other structures. As its name implies, finish carpentry is work that is going to be seen from the outside. Since this is going to be what people see, finish carpenters are expected to make their work as neat, clean and as finely-detailed as possible.

Rough and finish carpentry, however, are not the only types of carpentry. Other carpenters also specialize in different kinds of carpentry work. Cabinetmaking or cabinetry is a specialized branch of carpentry that deals with the making of stylish and durable cabinets for kitchen, homes and offices. Ship carpentry is another kind of carpentry that specializes in the construction, maintenance and repair of ships and boats. Another specialized field of carpentry is green carpentry. While this field requires the same set of skills as standard carpentry, it gives a lot of emphasis in utilizing environmentally-friendly materials and practices.

Carpentry work can also be categorized in terms of the kind of structure carpenters work on. Residential carpentry focuses on building new homes or remodeling old ones. The types of structures that residential carpenters construct include condominiums and townhomes. Commercial carpentry refers to carpentry used in the building of office buildings, malls and other commercial structures. These also include building and remodeling of schools, hospitals and hotels. Industrial carpentry, meanwhile, is the work done in industrial sites. Industrial carpenters assist in the building of dams, tunnel bracing and sewers.

Carpentry can be dangerous work. Since carpenters have to work with different types of equipment and tools, the risk of getting injured from cuts and wounds from improper handling of machines or from accidents exists. There is also the hazard posed by flying objects, fires, dust and toxic fumes as well as hearing risks because of the loud noise and vibration that mark construction sites. Moreover, carpenters are also expected to work high above the ground and as such the danger of falling also exists. To avert or minimize these hazards, carpenters must wear protective equipment and follow the safety protocols established by the company. They must also be knowledgeable about building codes to ensure that they will construct structures that are safe and durable for its occupants.

References

- What-is-insulation-types-of-insulation-845080: thebalancesmb.com, Retrieved 18 July 2018
- 5-most-common-thermal-insulation-materials: thermaxxjackets.com, Retrieved 29 June 2018
- Importance-architectural-lighting: tcpi.com, Retrieved 11 May 2018
- 3-important-elements-architectural-lighting: mullanlighting.com, Retrieved 28 March 2018
- Common-types-of-electrical-wiring-1152855: thespruce.com, Retrieved 19 June 2018
- Electronic-security-system: elprocus.com, Retrieved 11 June 2018

Permissions

Index